THE BEDFORD SERIES IN HISTORY AND CULTURE

Creating an American Culture, 1775–1800

A Brief History with Documents

Eve Kornfeld

San Diego State University

palgrave

CREATING AN AMERICAN CULTURE, 1775–1800: A BRIEF HISTORY WITH
DOCUMENTS by Eve Kornfeld

Library of Congress Control Number: 00-107181

Copyright Bedford/St. Martin's 2001
Softcover reprint of the hardcover 1st edition 2001 978-0-312-23702-8

PALGRAVE, 175 Fifth Avenue, New York, NY 10010

First published by PALGRAVE, 175 Fifth Avenue, New York, NY 10010. Companies
and representatives throughout the world. PALGRAVE is the new global imprint of
St. Martin's Press LLC Scholarly and Reference Division and Palgrave Publishers Ltd.
(formerly Macmillan Ltd.).

Manufactured in the United States of America.

6   5   4   3   2   1

f   e   d   c   b   a

ISBN 978-1-349-63132-2   ISBN 978-1-137-03834-0 (eBook)
DOI 10.1007/978-1-137-03834-0

TO MY FAMILY

Thomas Backer Simpson
Anna Kornfeld Simpson
Sara Kornfeld Simpson

Foreword

The Bedford Series in History and Culture is designed so that readers can study the past as historians do.

The historian's first task is finding the evidence. Documents, letters, memoirs, interviews, pictures, movies, novels, or poems can provide facts and clues. Then the historian questions and compares the sources. There is more to do than in a courtroom, for hearsay evidence is welcome, and the historian is usually looking for answers beyond act and motive. Different views of an event may be as important as a single verdict. How a story is told may yield as much information as what it says.

Along the way the historian seeks help from other historians and perhaps from specialists in other disciplines. Finally, it is time to write, to decide on an interpretation and how to arrange the evidence for readers.

Each book in this series contains an important historical document or group of documents, each document a witness from the past and open to interpretation in different ways. The documents are combined with some element of historical narrative—an introduction or a biographical essay, for example—that provides students with an analysis of the primary source material and important background information about the world in which it was produced.

Each book in the series focuses on a specific topic within a specific historical period. Each provides a basis for lively thought and discussion about several aspects of the topic and the historian's role. Each is short enough (and inexpensive enough) to be a reasonable one-week assignment in a college course. Whether as classroom or personal reading, each book in the series provides firsthand experience of the challenge — and fun—of discovering, re-creating, and interpreting the past.

Natalie Zemon Davis
Ernest R. May
David W. Blight

Preface

Embroiled in the "culture wars" of the early twenty-first century, Americans often assume that conflicts over our national identity and cultural diversity are new. Yet they are as old as our government itself. This volume offers some historical perspective for contemporary debates through an examination of the concerted efforts of American Revolutionary intellectuals to create a virtuous, independent, unified American culture. The founders' struggle to construct American identity reveals not only their intense desire for national unity but also the instability of national identities and the deep historical roots of American cultural diversity.

Keenly aware of the lack of unity that all but undermined the American Revolution, and convinced that the eyes of the world were on them, American intellectuals sought to overcome more than a century of regional, ethnic, and religious diversity in America. Their attempts to shape American ideas, values, and beliefs through a national culture took many forms. For some, a new, purer, and simpler American language seemed necessary to correct the corruptions and idiosyncrasies of English; for others, a national university and system of public education were prerequisites for a stable, virtuous society and polity. Still others tried to create an American history replete with national myths and heroic images in written or visual form. Waging losing battles against an emerging popular culture and the attractions of alternative cultures left these Revolutionary intellectuals scarred and weary by 1800. Yet their vision of American cultural unity, like the myth they created of Washington's cherry tree, has never entirely vanished, even in democratic, materialistic, pluralistic, modern America.

This book moves the subject of this quest for cultural unity from the margins to the center of historical inquiry. It invites students of history to pause in the usual rush from the Declaration of Independence through the Revolutionary War to the Constitutional Convention and the political battles of the 1790s. Is there another level to the

vii

story of the founding that this concentration on political events obscures? Should we not wonder why George Washington, Thomas Jefferson, Benjamin Franklin, and others cared so much about an American language, epic poetry, a national university, or American histories? Should we not try to understand the intellectuals' fear of American popular culture and the attractions of alternative, indigenous cultures?

To address these questions, this book crosses traditional disciplinary boundaries, much as the founders themselves did. If their efforts were multisensory, ours must be multidisciplinary. The interpretation that opens the book introduces and draws on a variety of approaches, methods, and perspectives, including social history, intellectual history, gender analysis, and poststructuralist theory. Its individual chapters can be read separately as coherent units or together to reveal the complexity and interconnectedness of the founders' intellectual project. Most important, the opening narrative attempts to open up the documents that follow to various readings. It invites readers to engage with the intellectuals' vision both sympathetically and critically and to emerge with a deeper understanding of the complicated, contested nature of national identities.

ACKNOWLEDGMENTS

This book has a long history. I first came upon republicanism and Philip Freneau while researching a junior paper and senior thesis at Princeton University. My fascination with the cultural history of the American Revolution led to a doctoral dissertation at Harvard University. Along the way, Dorothy Ross, John Murrin, Eric Foner, Maria DiBattista, Carl Schorske, John Clive, and Donald Fleming offered encouragement, freedom, guidance, and a model of a life of learning that inspires me to this day.

While teaching at Princeton and San Diego State University, my appreciation of American cultural diversity and my interest in poststructuralist theory and gender analysis grew. Fellowships from the National Endowment for the Humanities and the American Council of Learned Societies allowed me to frame the Revolutionary intellectual project in a new way. A Research, Scholarship, and Creative Activity Award and a sabbatical from SDSU provided support for a return to East Coast archives and time to write.

The responses of editors and readers of the *William and Mary Quarterly, Journal of the Early Republic, Canadian Review of American Studies*, and *Journal of American Culture* to my emerging interpretations enriched them immeasurably. So did conversations over the years with Mina Carson, William and Aimee Lee Cheek, Elizabeth Colwill, Robert Darnton, Natalie Davis, Cornelia Dayton, Edith Gelles, Douglas Greenberg, Jean Matthews, Michael Pratt, Jon Roberts, Emily and Norman Rosenberg, Francis Stites, and many, many of my students. I am profoundly grateful for the enthusiasm with which Charles Christensen of Bedford / St. Martin's greeted this manuscript and the care with which Joyce Appleby, Carol Berkin, Carl Prince, and Arthur Shaffer reviewed it. Lisa Moorehead and Rebecca Leyden offered the faith, support, and encouragement that come only from friends.

Without Tom Simpson, Anna Kornfeld Simpson, and Sara Kornfeld Simpson, this book might have been finished sooner, but my life would have been much poorer. As Anna regaled her first-grade class with tales of Mommy's book, two-year-old Sara checked my wastebasket daily for "her work" and merrily added her scribbling to mine. Sharing the cares and the joys of raising our children, Tom constantly demonstrated his commitment to my work as well as his own. Together, their appetite for new intellectual adventures is unsurpassed. This book is dedicated to my family, with all my heart.

Eve Kornfeld

Contents

PART ONE

Creating an American Culture

1

Introduction

"Americans, unshackle your minds and act like independent beings," Noah Webster urged his countrymen in 1788. "You have been children long enough, subject to the control and subservient to the interest of a haughty parent. You have now an interest of your own to augment and defend: you have an empire to raise and support by your exertions and a national character to establish and extend by your wisdom and virtues." To accomplish all of this, he argued, "Americans must *believe* and *act* from the belief that it is dishonorable to waste life in mimicking the follies of other nations and basking in the sunshine of foreign glory." Americans must create a national culture of their own.[1]

The irascible Connecticut "Schoolmaster to America" was accustomed to lecturing his readers and was unafraid to stand alone. But in this case, he enjoyed considerable support. His call for an independent American culture to form a "national character" or identity was echoed, less stridently but no less firmly, by many of the intellectual leaders of the Revolutionary generation. Whether their hopes centered on the creation of an American language, literature, education, or history, many American intellectuals believed that only a national culture could give Americans a sense of identity and unity.

To achieve these goals, a national culture would have to overcome more than a century of religious, ethnic, and regional diversity in America. The spectrum of settlement in seventeenth-century British America left the colonies with fundamentally different economic, social, and political systems and widely varying cultural values and beliefs. The holy experiments of New England and Pennsylvania rested uneasily on the same continent with the openly materialistic, exploitative colonies to the south. Contacts between the various British

[1]Noah Webster, *On the Education of Youth in America* (Boston, 1790), reprinted in *Essays on Education in the Early Republic,* ed. Frederick Rudolph (Cambridge, Mass., 1965), 77.

3

colonies in America were rare and sometimes contentious. The colonial elite were much more likely to travel "home" to England than to tour the other colonies.

This spectrum narrowed slightly in the eighteenth century, as all of the colonies were drawn more tightly into the British imperial system. Both British economic regulations and imported British culture muted colonial differences somewhat, as some colonists in different regions found common ground for the first time—if only through participation in Britain's Atlantic economy or imitation of British culture. The Great Awakening of evangelical Protestantism and the Enlightenment in America, both based on British cultural models, became the first significant intercolonial cultural movements. Far from creating a unified American culture, however, these two movements shared few values and beliefs. The awakened and enlightened tended to regard each other as misguided or even dangerous. The appearance of these two movements in the mid-eighteenth century marked the birth not of a national culture, but of a deep cultural division between evangelical enthusiasm and secular rationalism in America that has never been erased.

The coming of the Revolution did little to bridge these long-established cultural fissures. Indeed, they became more visible as the Revolutionary crisis demanded individual declarations of identity and loyalty. Even those colonists who decided to join the patriots did so with very different motives and expectations. Whereas colonial merchants and lawyers protested imperial restrictions on American trade and opportunities for the political participation and advancement of the colonial elite, urban artisans sought the redistribution of wealth and political power within America's seaport towns. In the countryside, the rural poor hoped to curb the power of local landlords and unresponsive colonial assemblies. The awakened joined the Revolution to restore moral order to their society, and the enlightened to permit rational progress. There was almost as little unity of purpose among the revolutionaries as between the patriots and loyalists in America.

The Revolutionary War exacerbated these internal tensions. States, communities, and even families were torn apart by the war. (In one famous example, Benjamin Franklin's illegitimate son William, the royal governor of New Jersey at the start of the Revolution, became a loyalist, causing his father great anguish.) Within states and communities, local issues and concerns often complicated the question of imperial loyalty. Religious, ethnic, class, and racial tensions, long simmering in the middle and southern colonies, bubbled to the surface during the war. The ferocity of the fighting within communities—to which both loyalist and patriot militias contributed through their campaigns of persuasion and coercion directed at civilians—made these differences of loyalty impossible to forgive or forget. Particularly in New Jersey and the Carolinas, the conventional war waged by the British and Continental armies was dwarfed by the terror of the civil war within communities. In New Jersey, arson and rape became tactics of war. In the Carolinas, the militias refused to give each other "quarter"—to accept surrender without death. Civilians in these areas quickly learned that neutrality was impossible in a revolutionary war.

Meanwhile, the Continental army was poorly supplied and neglected. The initial enthusiasm that had led many men to enlist and many citizens to provide them with high bounties and supplies quickly waned. One or two years into the war, General George Washington's urgent requests for payment and supplies for his troops went unheeded by the Continental Congress and the people it was powerless to tax. Civilians resisted the army's calls for blankets, wagons, and food. Most ominously, some Americans sought to profit from the war; some even sold bad meat and flour to the army. In the eyes of Washington and his officers, the American people seemed unwilling to sacrifice their individual self-interest for the public good. In their view, Benedict Arnold was not the only traitor to the patriot cause.

The Revolutionary turmoil also threatened the social order. In 1776, John Adams laughed off his wife Abigail's suggestion that he "Remember the Ladies" in America's new code of laws. But he laughed nervously. "As to your extraordinary Code of Laws, I cannot but laugh," he responded.

> We have been told that our Struggle has loosened the bands of Government everywhere. That Children and Apprentices were disobedient—that schools and Colleges were grown turbulent—that Indians slighted their Guardians and Negroes grew insolent to their Masters. But your Letter was the first Intimation that another Tribe more numerous and powerful than all the rest were grown discontented.—This is rather too coarse a Compliment but you are so saucy, I won't blot it out.[2]

---

[2]John Adams to Abigail Adams, April 14, 1776, reprinted in *The Book of Abigail and John: Selected Letters of the Adams Family, 1762–1784*, ed. L. H. Butterfield et al. (Cambridge, Mass., 1975), 122–23.

A few years later, Adams's seriocomic lament appeared quite prescient. Many traditional social relationships and hierarchies were disrupted during the war years. College students planted liberty trees, joined the army and militias, and pushed their parents and professors into more radical political positions. Apprentices were released to fight on one side or another, and servants fought side by side with their masters. Many class distinctions between white males temporarily dissolved in militias and Revolutionary councils.

In quieter social revolutions at home, women assumed men's accustomed roles, responsibilities, and authority as they managed households, families, farms, and businesses for extended periods of time while their husbands were absent. With growing confidence, they raised children, supervised servants and slaves, and made business decisions that they had formerly left to their husbands. Abigail Adams was but one among many such women. Women also acquired political ideas, commitments, and identities and were held accountable for them by militias and courts alike.

Even America's slaves found their situation altered during the war. In the northern states, some won their freedom by enlisting in the Continental army, with their masters' permission. Others, particularly in Virginia and Maryland, escaped slavery by joining the nearby British forces. Still others simply "stole themselves," melting away in the confusion of war-devastated South Carolina. Their mobility was especially frightening to their masters, of course, but it also alarmed other observers of this Revolutionary society. Although social disorder actually engulfed very few areas, some members of America's intellectual and political elite feared the imminent collapse of traditional social hierarchies and cultural authority in the wake of the Revolution.

Meanwhile, their active involvement in the Continental army and Congress gave many intellectuals a continental perspective—perhaps the first in American history. As they worked together during the Revolution, political and intellectual leaders from the northern, middle, and southern states began to perceive common interests and to speak in the same political language. They also shared a growing sense of revulsion at the rancor and divisiveness of American politics at all levels. Political disagreements among inexperienced elected officials sometimes exploded into violence on the floors of state legislatures, and factions emerged everywhere. Nor did the course of national politics run smoothly. The Confederation Congress was paralyzed by a pervasive distrust of strong central power, and the Constitutional Convention of 1787 crystallized continuing fears of centralization. As a party system desired by none emerged out of the Federalist-Antifederalist debates, the intellectuals' hopes for national political harmony evaporated.

The reemergence of local and regional antagonisms in the 1780s and 1790s reminded American intellectuals that a declaration of independence could not create a coherent nation. Shays's Rebellion of 1786, in which several thousand impoverished farmers in western Massachusetts closed down a county courthouse and marched on a federal arsenal, spurred political leaders to strengthen the national government. When farmers in western Pennsylvania displayed their disdain for federal taxation even with representation in the Whiskey Rebellion of 1794, the elite were convinced that the problem of national authority had not been solved by the adoption of the federal Constitution. Intense partisan political strife in the 1790s confirmed their deepest fears that without a national culture, the United States might fall apart before their very eyes.

The urgency seemed greatest to those intellectuals whose commitment to republicanism had led them into revolution. One of several competing ideologies in Revolutionary America, republicanism vied at times with popular evangelicalism, democratic radicalism, and a nascent liberal capitalism. Its appeal was particularly strong among the tiny minority of white males who were trained in the classics in colonial colleges from Harvard to William and Mary and who formed America's intellectual elite. They, more than most of their countrymen, were well prepared to shudder at the fragility of republics so richly illustrated in classical history and to apply history's lessons to the American Republic.

Based on English and European ideas reaching back to the Renaissance, American republicanism emphasized the civic nature of man. Only as active citizens in a self-governing republic could men achieve full humanity. To do so, they must learn to sacrifice their own interests and local interests for the common good, and they must be politically engaged and vigilant. Certain that those who were dependent on others could never develop civic virtue, republican theory limited citizenship to free white males with sufficient property to sustain their independence. But these citizens (the only fully human beings in the nation) were expected to act for the good of the whole society.

Although necessary to preserve a republic, civic virtue was extremely fragile. It was endangered not only by natural human laziness and selfishness but also by the tendency of governments to try to corrupt their citizens with wealth and privileges, or at least to lull them to sleep. According to republican theory, governmental power always expanded and encroached on individual liberty, once corruption overwhelmed civic virtue. Standing armies, established churches, patronage, and factionalism were sure signs to republicans of increasing governmental power and corruption. Together, they threatened each man's ability to develop his nature fully and freely in citizenship.

American republicans had watched in horror as these signs of corruption appeared in eighteenth-century England. It had moved some of them to revolt. Now they feared the reappearance of corruption in the American Republic. They knew of only one solution to this eternal problem. Only united, virtuous, liberty-loving citizens could resist corruption, curb governmental power, and prevent tyranny from overtaking the Republic. To survive, a republic needed citizens who would constantly put the public interest above their own. They must think first not of self, family, community, region, or party, but of the nation. Without a clear sense of national identity, citizens were even less likely to stave off self-interest, local pressures, and corruption. The creation of a strong, independent national culture seemed necessary to preserve the American Republic itself.

Thus many American intellectuals hoped to create a vital national culture to unify a heterogeneous society, to heal political divisions and quiet political contentiousness, to foster republican citizenship, and to achieve respect for the new state in the eyes of the world. Some attempted to invent a new American language and epic literature that would be purer and simpler than English. Others tried to design an American education for all citizens (and perhaps even for the future wives and mothers of citizens). Others turned to the construction of narratives of nationhood in the form of history, biography, visual images, and myths. Without national stories and heroes, they suggested, a nation could not cohere.

Defining American identity within a national culture also required intellectuals to declare who they were not. Imagining and excluding "the Other" became a central part of the intellectual project for some, who waged a losing battle against American popular culture or encountered alternative, indigenous cultures. Whereas some intellectuals emerged from these encounters ever surer of American identity and virtue, others found their confidence and their clarity of vision undermined. Yet their dream of American cultural unity, unrealized in their own time, never completely disappeared. Like the political system they designed, this dream became their legacy to future generations.

* * *

To place American intellectuals' efforts to create a national culture at the center of our inquiry is to enter the growing field of cultural history. Combining the close textual analysis characteristic of intellectual history with a concern for social context and dynamics drawn from social history, cultural historians have expanded our understanding of the formation and transformation of popular and elite ideas, values, attitudes, and beliefs. Over the past three decades, cultural historians have offered us richly drawn and carefully situated portraits of the cultural worlds of peasants, artisans, masters, slaves, shopkeepers, entrepreneurs, feminists, evangelicals, and intellectuals. They have explored popular *mentalité,* high culture, and the relationship between the two.

Absent from these studies are the allusions to national character, national identity, or "the American mind" that marked the work of earlier generations of scholars. Instead, recent studies have deepened our appreciation of the historical diversity and complexity of American culture. Taken together, they seem to suggest that a unified American mind or identity has never really existed, except as a myth or cultural construction. They invite us to ask why and how such myths were created, what social needs they might have served, and whom they silenced or made invisible.

This perspective, central to this and many other cultural histories, draws on the theory advanced by poststructuralist critics in a variety of disciplines that the construction of identity is a social act. Rather than growing naturally and inevitably, identities are often imagined, imposed, contested, and transformed by competing social groups, including intellectuals. Because the construction of identities involves the distribution of power, these cultural contests can be heated and even explosive. The most significant of these contests may reverberate through the life of a nation. If one social group or generation limits the definition of liberty, equality, or union, for example, another may revive the question and expand those boundaries. Thus no cultural debate is ever really closed, and no identity is final and complete.

To study intellectuals' attempts to create a national culture and identity, then, is not merely to learn about a small, closed circle of the elite. For while they act on their interests and reveal the limits of their visions, intellectuals also provoke questions, challenges, and alternative visions from those they address and those they ignore. The most isolated intellectuals, as withdrawn as Friedrich Nietzsche or Sigmund Freud, may disturb and unsettle their societies' cherished values and beliefs. Those as politically engaged as the American Revolutionary intellectuals—a rarity in world history—pose still more direct challenges to their contemporaries and posterity. Inscribed, contested, transformed, and reinscribed in narratives of nationhood, their ideas gradually acquire social power and perhaps even the status of self-evident truth. In success or in failure, their voices and visions echo through the centuries.

2

Inventing an American
Language and Literature

In the afternoon of September 25, 1771, two young graduates of the College of New Jersey (later called Princeton) delivered a stirring commencement poem titled "On the Rising Glory of America." The authors were Philip Freneau (soon to win fame as the "Poet of the Revolution" and later a central figure in the Jeffersonian Republican Party) and Hugh Henry Brackenridge (a lawyer, legislator, justice of Pennsylvania's supreme court, and author in later years). Their appreciative audience included John Witherspoon, the president of the college and a future signer of the Declaration of Independence; their classmate James Madison missed commencement due to poor health. The poem was so well received that it was published the following year in Philadelphia.[1] Its bold prediction that an American culture would soon arise to eclipse past European glories resonated with listeners at the commencement and readers at other colleges and in other colonies.

Opening with a strong declaration that it will treat "no more of Memphis," Alexandria, Greece, Rome, or even "Britain and her kings renown'd," the poem argues that although these empires made great contributions to civilization, they were spent. Drawing on the concept of *translatio imperii,* or the movement of the pinnacle (or imperial center) of civilization westward over the centuries from the Near East to Europe, the poem turns away from Britain's "illustrious senators, immortal bards, and wise philosophers" to look to the west.

A Theme more new, tho' not less noble, claims
Our cv'i y thought on this auspicious day;
The rising glory of this western world,
Where now the dawning light of science spreads
Her orient ray, and wakes the muse's song;

[1]Document 1

11

Where freedom holds her sacred standard high,
And commerce rolls her golden tides profuse
Of elegance and ev'ry joy of life.

As Britain's imperial sun was setting, the poem proclaims, another was rising in America.[2]

This expansive vision of "the rising glory of this western world" is quickly circumscribed. Disposing of the history of Spanish America in a few lines, Freneau and Brackenridge make clear that "America's own sons" lived in British America. "Better these northern realms deserve our song," the poem explains, "Discover'd by Britannia for her sons;/Undeluged with seas of Indian blood,/Which cruel Spain on southern regions spilt;/To gain by terrors what the gen'rous breast/Wins by fair treaty, conquers without blood."[3]

Following Britain's "bloodless" conquest, civilization gradually replaced savagery in North America. "Once on this spot perhaps a wigwam stood/With all its rude inhabitants, or round/Some mighty fire an hundred savage sons/Gambol'd by day, and filled the night with cries," completely ignorant of "fair science," agriculture, commerce, or "sacred truth." In poured the Europeans, moved by the purest idealism: "By persecution wrong'd/And popish cruelty, our fathers came/From Europe's shores to find this blest abode,/Secure from tyranny and hateful man." On "uncultivated tracts and wilds" they built towns, governments, and colleges. "Now view the prospect chang'd," the poem commands. "Learning exalts her head, the graces smile/And peace establish'd after horrid war/Improves the splendor of these early times."[4]

Improve the splendor it must, the poem suggests, for the logic of *translatio imperii* allowed no turning back. Once overwhelmed by corruption, "persecution," and "popish cruelty," the European empires could never be revived. "The dead, Acasto, are but empty names/And he who dy'd to day the same to us/As he who dy'd a thousand years ago," the young poets declare. The past was gone, and the future belonged to "America's own sons." If civilization was to survive and flourish, it must do so in America, for the westward movement of empire had reached its "final stage" and ultimate destination. "Hither [the Muses have] wing'd their way, the last, the best/Of countries

[2] Philip Freneau and Hugh Henry Brackenridge, *A Poem on the Rising Glory of America* (Philadelphia, 1772), reprinted in *The Poems of Philip Freneau,* ed. Fred L. Pattee (New York, 1963), 3 vols., 1:50–51.
[3] Ibid., 52.
[4] Ibid., 59, 61, 60.

where the arts shall rise and grow / Luxuriant, graceful . . . / This is a land of ev'ry joyous sound / Of liberty and life; sweet liberty! / Without whose aid the noblest genius fails, / And science irretrievably must die." Daring to search the "mystic scenes of dark futurity," the poem gazes unblinkingly at an American cultural millennium.[5]

The hopes expressed in Princeton in 1771 were widely shared among intellectuals of the Revolutionary generation. In the years that followed, many called on each other to "Prove to the world, in these new-dawning skies, / What genius kindles and what arts arise." At Yale's commencement of 1781, Joel Barlow urged his fellow graduates to make "blest Yalensia" the home of an American literature:

Enlarge her stores and bid her walls ascend;
Bid every art from that pure fountain flow,
All that the Muse can sing or man can know;
. . .
Say tis for them to stretch the liberal hand,
While war's dread tumults yet involve the land,
Sustain her drooping, rear her radiant eyes,
And bid her future fame begin to rise.[6]

Barlow's popular tutors at Yale College could not have agreed more. As early as 1770, Jonathan Trumbull had expected "Fair Freedom" to bless America with "splendor circling to the boundless skies" and to render it "the first in letters, as the first in arms." Timothy Dwight also waited for America to succeed "weak, doating [doting], fallen" England as the "queen of nations . . . / Scepter'd with arts, and arms." Logic demanded of the newly independent nation a national literature; the theory of *translatio imperii* further demanded that it become the culmination of Western civilization.[7]

That so many college graduates chose to cast their visions of American greatness in verse was no accident. Schooled in the classics, these men thought that the crowning glory of a national literature was its epic poetry. "As the Epic poem is the most noble of all others, the whole force of the human Genius is exhausted in beautifying it with

[5]Ibid., 63, 82, 71, 73.

[6]Jonathan Trumbull, "Lines Addressed to Messrs. Dwight and Barlow, on the projected publication of their Poems in London" (1775), in *The Poetical Works of John Trumbull* (Hartford, 1820), 109; Joel Barlow, *A Poem, spoken at the public commencement at Yale college, in New-Haven; September 12, 1781* (Hartford, [1781]), 6.

[7]Jonathan Trumbull, "Prospect of the Future Glory of America: Being the Conclusion of an Oration, delivered at the public commencement at Yale-College, September 12, 1770," in *Poetical Works*, 159; Timothy Dwight, *Greenfield Hill: A Poem in Seven Parts* (New York, 1794), 94–95.

figures, comparisons, and descriptions," Dwight explained in his Yale commencement address of 1772. "Nothing gives greater weight and dignity to Poetry, than Prophecy." Without this noblest and most difficult literary genre, American literature would be immature and incomplete, no worthy successor to Greek, Roman, or British literature.[8]

America's intellectual and political elite acted on this conviction throughout the early Republic. George Washington entrusted his legacy to America's epic bards, proclaiming them the keepers of "the gate by which Patriots, Sages and Heroes are admitted to immortality. Such are your Antient Bards who are both the priests and doorkeepers to the temple of fame." His name headed the subscription lists of America's first epic poems. Thomas Jefferson offered more substantial support to promising poets, as he recommended Joel Barlow and Philip Freneau to Washington for government appointments. Even John Adams, wary of many forms of cultural expression, urged Jonathan Trumbull to compose an epic poem in 1785: "I should hope to live to see our young America in Possession of an Heroick Poem, equal to those the most esteemed in any Country."[9]

The same year, Timothy Dwight attempted to reach "the height of arts." The Congregational minister and future president of Yale proudly presented his epic *The Conquest of Canäan; A Poem, in Eleven Books* as "the first of the kind, which has been published in this country."[10] Epic formula is followed closely throughout the poem, which begins with an invocation, presents an initial oration and reply, and dwells on detailed descriptions of evening and morning, camps and battles. The laments of fallen warriors are in grand epic language and style, and the ferocity of the single combat of heroes is lovingly described. Various visions and prophecies are scattered through the poem, with an extended "vision of futurity" reserved for the penultimate book.[11]

In his preface, Dwight apologizes for choosing "a subject, in which his countrymen had no national interest." His desire for "the fairest opportunities of exhibiting the agreeable, the novel, the moral, the pathetic, and the sublime" dictated his topic, he explains. Inspired by the modern epic poetry of the English bard John Milton, Dwight

[8]Timothy Dwight, *A Dissertation on the History, Eloquence, and Poetry of the Bible* (New Haven, Conn., 1772), 14, 16.

[9]Benjamin T. Spencer, *The Quest for American Nationality: An American Literary Campaign* (New York, 1957), 45, 58.

[10]Document 2

[11]Timothy Dwight, *The Conquest of Canäan; A Poem, in Eleven Books* (Hartford, 1785), title page, 1, 6, 13, 185–87, 4, 48.

sought to improve on the ancients with a biblical subject. The story of the Israelites' conquest of Canäan allowed him to celebrate divine justice and wisdom as no ancient poet could do. Moreover, as the Bible was "more elegant than *Cicero,* more grand than *Demosthenes,* . . . more correct and tender than *Virgil,* and infinitely more sublime than him who has long been honoured, not unjustly, with that magnificent appellation 'The Father of Poetry,'" Dwight could aspire to outdo them all. Even with a foreign subject, then, he hoped to throw "in his mite, for the advancement of the refined arts, on this side of the Atlantic."[12]

The most exalted character of the epic is the great leader of the Israelites. Joshua, "a Chief divine! with every virtue crown'd,/In combat glorious, and in peace renown'd," exhibited a wise, paternal concern for his people, as well as unmatched bravery and strength in war. The poem begins with a summary of his achievements and a dedication to "the Chief, whose arm to Israel's chosen band/Gave the fair empire of the promis'd land,/Ordain'd by Heaven to hold the sacred sway." Yet the epic is also dedicated "To his Excellency, GEORGE WASHINGTON, Esquire, Commander in Chief of the American Armies. The Saviour of his Country, The Supporter of Freedom, And the Benefactor of Mankind." Dwight's readers could hardly have missed the parallel or the veiled references to America as the new Israel, the "fair empire of the promis'd land," divinely ordained to preserve sacred virtue. The poet's prefatory apology notwithstanding, the subject of this first American epic was, at least symbolically, American.[13]

The poem's portrait of a just war reinforces this symbolic connection. So does the end of the epic, where both a tired narrative and the distress of war are relieved by a sustained prophecy of Israel's divine mission. This vision of "long futurity" reaches even beyond Milton in employing the prophetic device to recount the rest of biblical history. At last making explicit the parallel between the original and the new Israel, Dwight brings Joshua's eyes to rest on "a new Canäan's promis'd shores": "Lo, there a mighty realm, by heaven design'd/The last retreat for poor, oppress'd mankind!"[14]

Overcoming and redeeming the "slavery of the eastern Continent," America provided the scene for the culmination of the vision. The "Glory of the Western Millennium," or the triumph of Christianity on earth at the end of human history, emerged on American shores.

[12]Dwight, *Dissertation,* 3–4.
[13]Dwight, *Conquest,* 49, 2, 13, 1.
[14]Ibid., 59–63, 235, 254, 253.

"Here Empire's last, and brightest throne shall rise," the poem concludes, "And Peace, and Right, and Freedom, greet the skies." By inserting American glory in an unlikely biblical context, Dwight constitutes America as an epic subject and links it to the coming of the earthly millennium. Then he closes his poem, leaving the full praise of American virtue for other bards to "sing, to earth's remotest clime."[15]

Accepting this challenge to carry cultural translation a step further was Dwight's friend and student at Yale, Joel Barlow. Only a year after graduating from college, Barlow began to design an epic poem with Columbus as his hero. He wished to acquaint Americans "with the life and character of that great man, whose extraordinary genius led him to the discovery of the continent, and whose singular sufferings ought to excite the indignation of the world." He was convinced that "every circumstance relating to the discovery and settlement of America" ought to hold intrinsic interest for his fellow citizens, who knew all too little about Columbus. Yet Barlow quickly realized that he wanted the focus of the poem to be on America itself, "the most brilliant subject" possible, and not on Columbus's heroic struggles.[16]

The result was a significant departure from conventional epic form. The story of Columbus's efforts, patience, and paternal wisdom in governing rebellious sailors (an implicit comparison with both Joshua and Washington) is collapsed into Barlow's introduction. The vision that traditionally closes an epic and consoles the hero "by unfolding to him the importance of his discoveries, in their extensive influence upon the interest and happiness of mankind, in the progress of society" expands to take over the poem completely. *The Vision of Columbus; A Poem in Nine Books*,[17] finally published in 1787, combines Freneau and Brackenridge's prophecy of American greatness with Dwight's epic length and scope.[18]

The Vision of Columbus thus became the first American epic to deal directly with our "own Canäan." The poem begins with an exuberant description of American nature, wild and sublime, and infinitely superior to European scenes: "And hills unnumber'd rose without a name, / Which placed, in pomp, on any eastern shore, / Taurus would shrink, the Alps be sung no more; / For here great nature, more exalted show'd / The last ascending footsteps of her God." Extravagant

[15]Ibid., 255, 168.
[16]Joel Barlow, *The Vision of Columbus; A Poem in Nine Books* (Hartford, 1787), "Introduction," vii, xxi.
[17]Document 3
[18]Ibid., xiv, xx.

praise and epic language are supplemented by the ample use of American place names.

> Now the still morn had tinged the mountain's brow
> And rising radiance warm'd the plains below;
> Stretch'd o'er Virginian hills, in long array,
> The beauteous Alleganies met the day.
> From sultry Mobile's rich Floridian shore,
> To where Ontario bids hoarse Laurence roar,
> O'er the clear mountain tops and winding streams,
> Rose a pure azure, streak'd with orient beams;
> Fair spread the scene, the hero gazed sublime,
> And thus in prospect hail'd the happy clime.[19]

Still more lavish attention is paid to American history, particularly as the vision is "confined to North America" midway through the poem. American independence becomes the centerpiece of Columbus's vision, and the poem grandly records the names and deeds of America's military and political heroes. Epic forms are most evident in these central chapters, as the valiant orations of ancient and biblical military leaders echo in General Washington's speeches to his American army:

> Now, the broad field as gathering squadrons shade,
> The sun's glad beam their shining ranks display'd;
> The glorious leader waved his glittering steel,
> Bade the long train in circling order wheel;
> And while the banner'd hosts around him roll,
> Thus into thousands speaks the warrior's soul:
> Ye patriot chiefs, and every daring band,
> That lift the steel or tread the invaded strand,
> Behold the task! these beauteous realms to save,
> Or yield whose nations to an instant grave.
>   . . .
> Rise then to war, to noble vengeance rise,
> Ere the grey sire, the helpless infant dies;
> Look thro' the world, where endless years descend,
> What realms, what ages on your arms depend!
> Reserve the fate, avenge the insulted sky;
> Move to the strife, we conquer or we die.[20]

American heroism in battle, however glorious, forms only a part of *The Vision of Columbus.* Moving well beyond the conventions of the

---

[19]Ibid., 30, 34, 35.
[20]Ibid., 152, 157–62, 164–65, 170–71.

genre, this American epic also celebrates the potential of American agriculture and commerce to realize Columbus's vision of world peace and harmony. Yet the poet's highest hopes are reserved for American culture. Instructional and inspirational, art—and particularly epic poetry—must rekindle and spread the spirit of Columbus: "So, while imperial Homer tunes the lyre,/The living lays unnumber'd bards inspire,/From realm to realm, the kindling spirit flies,/Sounds thro' the earth and echoes to the skies." Only the combined strength of American "arts and laws" could finally bind mankind "by leagues of peace." American artists, scientists, and poets, "the boast of genius and the pride of song," would offer "new guidance to the paths of man." Above all, the epic poetry of "blaring Dwight" (and trumpeting Barlow?) served the nation and the world and "revives the promis'd land."[21]

Unlike Dwight's *Conquest of Canäan, The Vision of Columbus* refuses to leap from American glory to the Christian millennium. Columbus's angelic guide insists that a glimpse of the millennium would blind him. He must content himself with this world, transformed by American law, commerce, and literature. The divine "progressive plan" will unfold, the Angel assures Columbus, through the cooperative (secular) efforts of bold political and commercial "chiefs like thee, with persevering soul," and the equally daring poets of America.

> Then, while the daring Muse, from heavenly quires,
> With life divine the raptured bard inspires,
> With bolder hand he strikes the trembling string,
> Virtues and loves and deeds like thine to sing.
> No more with vengeful chiefs and furious gods,
> Old Ocean crimsons and Olympus nods,
> Nor heavens, convulsive, rend the dark profound,
> Nor Titans groan beneath the heaving ground;
> But milder themes shall wake the peaceful song,
> Life in the soul and rapture on the tongue;
> To moral beauties bid the world attend,
> And distant lands their social ties extend,
> Thro' union'd realms the rate of conquest cease,
> War sink in night, and nature smile in peace.

Thus the poem ends on a secular millennial vision of universal peace and harmony created by commerce and culture in America.[22]

[21]Ibid., 138, 146, 203–4, 207–12, 211.
[22]Ibid., 214, 226, 229, 222–24, 247, 241, 224.

To this goal Barlow hoped to contribute, not only through his epic's moral lessons but also through its language. Both, he claimed, were purer than those of Homer and Virgil, whose celebrated epics inflamed the soul for war and preached the divine right of kings. Their martial language supported these ancient poems' pernicious political lessons, Barlow believed. His own epic linguistic goal was to move the world from a confused multitude of languages to a universal tongue. Indeed, the millennial consummation of *The Vision of Columbus* depends largely on the adoption of a universal "pure language," to ease the communication of scientific, artistic, political, and moral truths among all peoples. Barlow tried to advance this goal by introducing American names and expressions into his epic whenever possible. To conflate American idioms and a universal language appears to have been no more problematic to Barlow than to locate the millennial hopes of the entire world in the new nation.[23]

Barlow's commitment to this vision remained strong throughout his life. Even as his unusual career as a U.S. diplomat carried him to Algiers, France, and Poland (where he died in hot pursuit of Napoleon in 1812), he continued to revise and expand his epic poem. His *Columbiad* of 1807 incorporated still more American names and idioms, all in the hope of creating a pure, universal language that could help erase national divisions and misunderstandings.

Barlow owed his linguistic vision to his Yale classmate and friend, Noah Webster. The son of a modest Connecticut farmer who mortgaged his farm to send his son to Yale, Webster became a teacher, lawyer, political leader, and author after his graduation in 1778. He also became America's first lexicographer, devoting years of his life to studying and lecturing about American pronunciation, spelling, and grammar. His speller of 1783, renamed *The American Spelling Book* in 1788, was the first and most widely used American textbook of the period; it sold more than 24 million copies in its first half century. His *Compendious Dictionary* of 1806 and massive, unabridged dictionary of 1828 earned Webster lasting fame and respect. Few of his fellow intellectuals could claim as much success in shaping American culture.

Yet as generations of American children used his "Blue-Backed Speller" and their parents told each other to "look it up in Webster's," few realized that these books were actually designed as testaments to American linguistic superiority. For a brief but intense period in the

---

[23]Ibid., 250, 252.

1780s, Webster attempted to create an American language whose purity and simplicity would reflect and sustain American virtue and independence. American independence would not be complete, he believed, until Americans began to speak and write in a language of their own.

As early as 1783, Webster began to argue that American political independence depended on cultural independence and purity. "The author wishes to promote the honour and prosperity of the confederated republics of America; and chearfully throws his mite into the common treasure of patriotic exertions," the introduction to his famous speller announced. "This country must in some future time, be as distinguished by the superiority of her literary improvements, as she is already by the liberality of her civil and ecclesiastical constitutions." Only by staving off the influence of Europe, "grown old in folly, corruption and tyranny," and by judging Europe's "perverted" laws, "licentious" manners, "declining" literature, and "debased" human nature for what they were, could America achieve its rightful position in the world. With youthful exuberance, Webster concludes,

> It is the business of *Americans* to select the wisdom of all nations, as the basis of her constitutions,—to avoid their errours,—to prevent the introduction of foreign vices and corruptions and check the career of her own—to promote virtue and patriotism,—to embellish and improve the sciences,—to diffuse an uniformity and purity of *language,*—to add superiour dignity to this infant Empire and to human nature.[24]

Yet the speller itself took only small, measured steps toward this broad goal. Its lists of spelling words include the names of American national heroes and places suitable for "lisping" at an early age to instill national pride. A table of "lessons of easy words to learn [sic] children to read and to know their duty" begins with simple moral and religious warnings that "sin will lead us to pain and woe," and graduates to more complex lessons in honesty, obedience, industry, and patriotism. All of these lessons are then interwoven in a narrative of a virtuous and a wicked brother, the "Story of Tommy and Harry." But as several words (honour, chearfully, errours, superiour) in Webster's plea for cultural independence above make clear, the language and orthography of the speller remained undeniably English.[25]

[24]Noah Webster, *A Grammatical Institute of the English Language* (Hartford, 1783), 1:14–15.
[25]Ibid., 101–3, 106–13.

Still searching for a more striking way to declare American linguistic independence in 1786, Webster began to correspond with Benjamin Franklin. Then eighty years old, Franklin had long advocated a radical reform of the alphabet to simplify and regularize English spelling. Franklin's scheme, which introduced many new characters to convey pure sounds, had attracted little support over the years.[26] He now saw an opportunity to convert a young, energetic Webster to the cause of radical linguistic reform. Franklin's constant encouragement and support between 1786 and his death in 1790 both flattered and convinced Webster. Although his plan for changes in the alphabet was never as radical as Franklin wished, Webster was soon proclaiming orthographic reform both practical and necessary.

Characteristically, Webster linked reform in spelling to the creation of an American language. "A national language is a national tie, and what country wants it more than America?" he asked a sympathetic correspondent in 1786. As he explained at the first meeting of the American Philological Society, which he founded in New York in 1788, a simple, clear American language was urgently needed to preserve American virtue and perhaps Christianity itself. "In the infancy of an institution, founded for the particular purpose of ascertaining and improving the American tongue, it may be useful to examine the importance of the design, and show how far truth and accuracy of thinking are concerned in a clear understanding of *words*," Webster declared. "If it can be proved that the *mere use of words* has led nations into error, and still continues the delusion, we cannot hesitate a moment to conclude, that grammatical enquiries are worthy of the labor of *men*."[27]

In his *Dissertations on the English Language: with Notes, Historical and Critical* of 1789, dedicated to Benjamin Franklin, Webster developed his linguistic theory more fully.[28] The text first attempts to establish the superiority of the English language (like the English constitution) over those of other European nations:

> The spoken language is also softened, by an omission of the harsh
> and guttural sounds which originally belonged to the language,
> and which are still retained by the Germans, Scotch and Dutch.
> At the same time, it is not, like the French, enervated by a loss of

[26]Document 4

[27]Noah Webster to Timothy Pickering, New York, May 25, 1786, in *Letters of Noah Webster*, ed. Harry R. Warfel (New York, 1953), 52; Noah Webster, *The American Magazine*, May 1788, 399, 400–401.

[28]Document 5

consonants. It holds a mean between the harshness of the German, and the feebleness of the French. . . . As the English have attempted every branch of science, and generally proceeded farther in their improvements than other nations, so their language is proportionably copious and expressive.[29]

Yet the English language was not secure in its position of purity and superiority. Due to the imperfection of man, a language cannot improve forever. "But when a language has arrived at a certain stage of improvement, it must be stationary or become retrograde; for improvements in science either cease, or become slow and too inconsiderable to affect materially the tone of a language." The golden age of the English language in England lay in the past, according to Webster, beginning with the reign of Queen Elizabeth and ending (significantly) with that of George II. This period was indeed glorious, marked by an abundance of "writers of the first class." Webster laments, "It would have been fortunate for the language, had the stile of writing and the pronunciation of words been fixed, as they stood in the reign of Queen Ann[e] and her successor."[30]

Such was not the case. Time took its toll on the English and their language: "Few improvements have been made since that time; but innumerable corruptions in pronunciation have been introduced by Garrick, and in stile, by Johnson, Gibbon and their imitators." This was no accidental development; the English people must take full responsibility. Webster concludes, "Such however is the taste of the age; simplicity of stile is neglected for ornament, and sense is sacrificed to sound." As the English people relaxed their historical activity, the corruption of the once pure English language was inevitable. The traditional simplicity and sense of the English people no longer set the standards of English pronunciation. Instead, those English authors "who have attempted to give us a standard, make the practice of the court and stage in London the sole criterion of propriety in speaking." Where could vice, caprice, and corruption be found in more abundance than the English court and stage? All, it seemed, was lost.[31]

In a logical, if extremely judgmental, conclusion, Webster proclaims English linguistic corruption both inevitable and incurable. "The corruption however has taken such deep root in England, that there is little probability it will ever be eradicated," he writes. "Such is the

[29]Noah Webster, *Dissertations on the English Language: with Notes, Historical and Critical* (Boston, 1789), 60.
[30]Ibid., 30.
[31]Ibid., 30, 34, 24–25.

force of custom, in a nation where all fashionable people are drawn to a point, that the current of opinion is irresistable; individuals must fall into the stream and be borne away by its violence." Drawing on Revolutionary political thought for parallels, Webster locates the ultimate cause of English linguistic corruption in the structure of society, in which "all fashionable people are drawn to a point" at the royal court. This evil is beyond the help of a mere grammarian.[32]

Irresistibly, then, Webster turned to America for the preservation and enhancement of the English language. "On examining the language, and comparing the practice of speaking among the yeomanry of this country, with the stile of Shakespear and Addison," he declares, "I am constrained to declare that the people of America, in particular the English descendants, speak the most pure *English* now known in the world." The American yeoman farmer had become the repository not only for the political virtue but also for the linguistic purity of the golden age of England. This was due primarily to his isolation from "pretended refiners of the language" in England, who acted "from mere affectation." The *Dissertations* are full of examples of the differences between the American and English styles of speech, about which Webster becomes impassioned:

> *WRATH,* the English pronounce with the third sound of *a* or *aw;* but the Americans almost universally preserve the analogous sound, as in *bath, path.* This is the correct pronunciation; and why should we reject it for *wroth,* which is a corruption? If the English practice is erroneous, let it remain so; we have no concern with it: By adhering to our own practice, we preserve a superiority over the English, in those instances, in which ours is guided by rules; and so far ought we to be from conforming to their practice, that they ought rather to conform to ours.[33]

Thus the *Dissertations* are Webster's declaration of American cultural independence. "We have therefore the fairest opportunity of establishing a national language, and of giving it uniformity and perspicuity, in North America, that ever presented itself to mankind," he declares. Americans must resist the temptation to "adopt promiscuously" English tastes, opinions, and manners. "Customs, habits, and *language,* as well as government should be national. America should have her *own* distinct from all the world," he urges. "To copy foreign manners implicitly, is to reverse the order of things, and begin our

[32] Ibid., 176.
[33] Ibid., 288, 124.

political existence with the corruptions and vices which have marked the declining glories of other republics." Before foreign "corruptions and vices" overwhelmed American linguistic purity, Americans must develop a national language of their own.[34]

Webster's orthographic scheme itself consisted of three principal alterations. First, he would omit silent characters, leaving out such letters as the *a* in *bread* as useless and confusing. Second, he would substitute letters with a definite sound for those with a vague or indeterminate pronunciation, so that *ea* or *ie* would become *ee*. Third, he would make a small alteration in a character to distinguish its different sounds, putting a stroke across *th* or a dot over *a* to signify a change in pronunciation. Pursuing throughout a method of purifying and simplifying by analogy and in accord with common usage, Webster also wished to abandon the subjunctive as an unnatural and unnecessary sophistication added to older and more common modes. Insisting always that he was returning to older usage rather than innovating (echoing his fellow revolutionaries' political claims), Webster attempted to halt the decaying process of time by constructing "a perfect correspondence between spelling and pronunciation."[35]

Through these modest reforms, Webster hoped to create a uniform national language and "to reconcile the people of America to each other, and weaken the prejudices which oppose a cordial union." If Webster's vision was more national and less universal than that of Barlow, it was no less comprehensive. For it was essentially a linguistic millennium that Webster sought, with its accompanying effects of peace and harmony throughout the new nation.[36]

By 1790, Webster was ready to model this American language for his dubious readers. His *Collection of Essays and Fugitiv Writings* of 1790 both incorporates and defends his proposed orthographic reforms:[37]

> In the essays, ritten within the last yeer, a considerable change of spelling iz introduced by way of experiment. This liberty waz taken by the riters before the age of queen Elizabeth, and to this we are indeted for the preference of modern spelling, over that of Gower and Chaucer. The man who admits that the change of *housbonde, mynde, ygone, moneth* into *husband, mind, gone, month*, iz an improovment, must acknowledge also the riting of *helth, breth, rong,*

[34]Ibid., 36, 179.
[35]Ibid., 35.
[36]Ibid., 36.
[37]Document 6

tung, munth, to be an improovment. There iz no alternativ. Every possible reezon that could ever be offered for altering the spelling of words, stil exists in full force; and if a gradual reform should not be made in our language, it wil proov that we are less under the influence of reezon than our ancestors.[38]

Besieged by ridicule and complaints that such spelling changes would soon render older English books incomprehensible to Americans, Webster replied that this result might be the most important advantage of his reform. Americans read and copied too many English books and ideas, he believed. "Every engine should be employed to render the people of this country *national;* to call their attachments home to their own country; and to inspire them with the pride of national character," he argues. "However they may boast of Independence, and the freedom of their government, yet their *opinions* are not sufficiently independent; an astonishing respect for the arts and literature of their parent country, and a blind imitation of its manners, are still prevalent among the Americans." True independence required as clear a distinction between national cultures and characters as existed between political institutions, Webster insisted. A primary advantage of his scheme, then, "would be, that it would make a difference between the English orthography and the American" and "encourage the publication of books in our own country," a crucial step in the development of an independent American culture.[39]

Few of his countrymen agreed. Forced to abandon the more radical of his orthographic reforms due to adverse public reactions and the loss of Franklin's invaluable support in 1790, Webster scaled back his spelling changes in subsequent books. Yet he never ceased to labor and hope for an American national language and cultural independence. Never would he bow to corrupt English linguistic customs. He continued to base his rules for pronunciation on that of "the body of the people," insisting always that the only true standards were "the rules of the language itself, and the general practice of the nation."[40]

Eager as he was to locate such a unifying "general practice" in America, Webster encountered only a series of local dialects and prejudices. As he traveled around the country to promote his textbooks and linguistic plan, he was surprised to find his own pronunciation and manners ridiculed. "Nothing can be more illiberal than the prejudices

---

[38]Noah Webster, *A Collection of Essays and Fugitiv Writings; on Moral, Historical, Political and Literary Subjects* (Boston, 1790), xi.
[39]Webster, *Dissertations,* 397–98.
[40]Ibid., 152, 27–28.

of the southern states against New-England manners," he noted. "They deride our manners and by that derision betray the want of manners themselves. However different may be the customs and fashions of different states; yet those of the southern are as ridiculous as those of the northern. The fact is, neither one nor the other are the subjects of ridicule and contempt." The outcome of Webster's search for "the general practice of the nation" can be read clearly in this passage.[41]

He was still more explicit in a private letter of 1793 to a Yale classmate and fellow New Englander. "I was at war with men and things which did not come up to my ideas of New England excellence. I fought many hard battles in defence of Connecticut," he confessed. "My partiality for the institutions which gave birth to and still support our manners will cease only with life. New England is certainly a phenomenon in civil and political establishments, and in my opinion not only young gentlemen from our sister states, but from every quarter of the globe, would do well to pass a few years of their life among us and acquire our habits of thinking and living."[42]

Webster's desire to send the nation to school in New England is apparent in all of his works. From the first speller to the last dictionary, his rules for spelling, grammar, and pronunciation reflect provincial New England usage. "Ask any plain countryman, whose pronunciation has not been exposed to corruption by mingling with foreigners, how he pronounces the letters, *t, r, u, th,* and he will not sound *u* like *eu,* nor *oo,* but will express the real primitive English *u,*" he insisted. Yet only Yankees spoke this way. The "general practice" that formed Webster's standard was that of New England. Never really repudiating his early desire to make American spelling, pronunciation, and usage uniform and national, he also never recognized how imperial this gesture was. For the American language that Noah Webster sought to create was native only to New England.[43]

[41]Noah Webster, *Sketches of American Policy* (Hartford, 1785), 44–45.
[42]Noah Webster to Oliver Wolcott, Hartford, May 25, 1793, in *Letters,* 110–11.
[43]Webster, *Dissertations,* 152.

3

Educating American Citizens

When Benjamin Rush called in 1786 for an American education that would transform young men into "republican machines," he articulated both a utopian vision and an actual intellectual project.[1] A 1760 graduate of the College of New Jersey, Rush was a statesman, physician, and professor of medicine in Philadelphia until his death in 1813. He also served as a friend, mentor, correspondent, and conscience to many of his former classmates, colleagues, and students. Few issues of his day escaped his attention. Yet none engaged him more thoroughly and consistently than the need to create a unified system of republican education in America.[2]

An extensive network of the American intellectual elite shared Rush's judgment that the education of youth must form the keystone of a new American culture. In a widespread and intense exchange on the subject of republican education in the 1780s and 1790s, American intellectuals launched a campaign to replace the existing haphazard mixture of pay, charity, and district schools with a uniform system of public education. Their numerous pamphlets, books, speeches, and bills were intended to inspire legislative action, to fire public enthusiasm for the execution of the schemes, and to formulate the contours of an education that could meet the needs of the new nation.

Among the leading educational thinkers and reformers of the early Republic were Thomas Jefferson, Noah Webster, Robert Coram (editor of the *Delaware Gazette*), Samuel Knox (a Maryland minister), Samuel Harrison Smith (a Pennsylvania journalist), and Rush himself. None of these men believed that education was limited to the classroom; their own experiences proved otherwise. Yet all of them placed special emphasis on the creation of a formal republican education in a

---

[1] Document 7
[2] Benjamin Rush, *A Plan for the Establishment of Public Schools and the Diffusion of Knowledge in Pennsylvania; to Which are Added, Thoughts Upon the Mode of Education Proper in a Republic* (Philadelphia, 1786), 27.

uniform system of public schools. The private tutors, boys' academies, and dame schools scattered across the country seemed unlikely to create the unswerving devotion to republicanism that these intellectuals considered essential to America's survival. Although their particular plans differed on certain points, together they articulated a fundamental, coherent vision of republican education.

This republican vision contained a complicated mixture of liberating and restraining elements. First and foremost, this was to be an education for citizenship. Dedicated to creating a large body of citizens capable of guarding their liberties jealously against the ever-encroaching powers of government, the reformers sought to spread literacy and basic civic knowledge as widely as possible among American citizens. The common man must be educated to defend himself and the Republic against corruption. Drawing on the cognitive theory of the English philosopher John Locke, American intellectuals regarded children's minds as "blank slates" that must be filled in properly if the Republic was to succeed.

Thus Thomas Jefferson wrote from Paris in 1786 to a fellow Virginian that "by far the most important bill" in their state's proposed code of laws was "that for the diffusion of knowledge among the people. No other sure foundation can be devised for the preservation of freedom, and happiness." As the Revolutionary governor of Virginia, Jefferson himself had drafted the bill, which included a comprehensive plan of public education from elementary schools to a university. Now, surrounded by the "ignorance, superstition, poverty and oppression in body and mind" of the European masses, Jefferson urged his correspondent to greater exertion: "Preach, my dear sir, a crusade against ignorance; establish and improve the law for educating the common people."[3]

Jefferson was not alone. In the face of public and legislative indifference or hostility, all of the republican reformers emphasized the need for a broad system of public schools to offer, at least, a primary education to all citizens. Although few of his fellow intellectuals employed the inflamed rhetoric of Robert Coram, all shared his expansive vision of "universal" (white male) elementary education.

> Education then ought to be secured by government to every class of citizens, to every child in the state. . . . Education should not be left

[3]Thomas Jefferson to George Wythe, Paris, August 13, 1786, in *The Papers of Thomas Jefferson,* ed. Julian P. Boyd (Princeton, N.J., 1950–), 10:243–45. The bill was defeated in the Virginia state legislature.

to the caprice, or negligence of parents, to chance, or confined to
the children of wealthy citizens: it is a shame, a scandal to civilized
society, that part only of the citizens should be sent to colleges and
universities to learn to cheat the rest of their liberties.[4]

Coram hoped that free, compulsory, universal primary education
would eliminate poverty, which entailed "degradation and distress"
and threatened the "security of all governments." In promoting literacy and fundamental civic knowledge for all American citizens, the
republican educational vision was clearly liberating and far more egalitarian than the prevailing American (or European) educational institutions.[5]

Yet this represented only part of the reformers' vision. None of the
republican educational theorists believed that literacy and "a diffusion
of knowledge" alone could create a republican citizenry. The other
essential ingredient—indeed, the most essential ingredient—was
civic virtue, or the willingness to sacrifice self-interest for the public
good. Essay after essay echoed Rush's call for a systematic, deliberate
cultivation of this republican virtue.

Let our pupil be taught that he does not belong to himself, but that
he is public property. Let him be taught to love his family, but let
him be taught at the same time that he must forsake and even forget them when the welfare of his country requires it. . . . He must
love private life, but he must decline no station, however public or
responsible it may be, when called to it by the suffrages of his fellow
citizens. . . . Above all he must love life and endeavour to acquire as
many of its conveniences as possible by industry and economy, but
he must be taught that this life "is not his own," when the safety of
his country requires it.[6]

Rush believed this sort of education "to be peculiarly necessary in
Pennsylvania, while our citizens are composed of the natives of so

[4] Robert Coram, *Political Inquiries: to which is Added, a Plan for the General Establishment of Schools throughout the United States* (Wilmington, Del., 1791), reprinted in
Essays on Education in the Early Republic, ed. Frederick Rudolph (Cambridge, Mass.,
1965), 79–145, quotation on 113.
[5] Ibid., 123. Coram and the other republican educational reformers were disappointed. The state legislatures of the early Republic, although constitutionally enabled
and directed to promote education, produced few plans or laws and successfully executed even fewer. The only serious legislative action to establish a system of state education, taken in New York in 1795 at the prompting of Governor George Clinton,
allocated twenty thousand pounds to found more than one thousand schools. This legislation was allowed to expire in 1800, and the allocation was withdrawn.
[6] Rush, *Plan,* 20–22.

many different kingdoms in Europe. Schools of learning, by producing one general and uniform system of education, will render the mass of the people more homogeneous and thereby fit them more easily for uniform and peaceable government." He shared Jefferson's fears that continued immigration would make American society "a heterogeneous, incoherent, distracted mass," but Rush preferred civic education to restrictions on immigration. Any European immigrant could become an American, he believed, with the proper education.[7]

Like George Washington, Rush worried about "those local prejudices and habitual jealousies . . . which, when carried to excess, are never failing sources of disquietude to the Public mind, and pregnant of mischievous consequences to this Country." This worry led both men to advocate the establishment of a national university to cap a system of public education.[8] There the most talented youths could mingle and overcome those local allegiances and prejudices fostered in families and communities. As Washington explained, "Among the motives to such an institution, the assimilation of the principles, opinions, and manners of our youth from every quarter, will deserve attention. The more homogeneous our citizens can be made in these particulars, the greater will be our prospect of permanent union."[9]

During the federal Constitutional Convention of 1787, Washington, Benjamin Franklin, and James Madison proposed to establish a national institution for the advancement of the arts and sciences. They were dissuaded only by the conviction of the convention that the federal government certainly would possess the power to create one. As president, Washington formally (and futilely) made the request in his first message to Congress in 1790 and his last in 1796. He then bequeathed a small endowment to establish a national university in his Last Will and Testament of 1799.[10]

Jefferson echoed Washington's appeal in his own presidential messages to Congress in 1806 and 1808. He hoped to build a national university around Charles Willson Peale's Philadelphia museum of art, natural history, and technology, which attracted more than ten thou-

---

[7]Ibid., 14; Thomas Jefferson, *Notes on the State of Virginia* (Philadelphia, 1788), 93.
[8]Document 8
[9]George Washington, Last Will and Testament, in *The Writings of George Washington from the Original Manuscript Sources, 1745–1799*, ed. John C. Fitzpatrick (Washington, D.C., 1940), 37:279–81. *Annals of Congress*, 4th Cong., 2nd sess., 1519, cited in Ellwood P. Cubberley, *Readings in Public Education in the United States* (Boston, 1934), 243.
[10]Document 9

sand visitors each year to view its carefully preserved, ordered, and classified "world in miniature."[11] Ever since opening the museum's doors during the Constitutional Convention of 1787, Peale had hoped to attract governmental support to expand his exhibits and make his a truly *National Museum.*" He wrote to Jefferson in 1802 that such an institution would be "more powerful to humanize the mind, promote harmony, and aid virtue, than any other School yet imagined."[12] As these efforts failed to rouse Congress to action, the republican reformers continued to stress the need for a moral education (in schools rather than in families) that would restrain natural selfishness and local attachments and systematically inculcate patriotism, self-control, and civic virtue.

A decade after Rush's pioneering essay, remarkably similar educational goals were articulated by Samuel Harrison Smith and Samuel Knox. They shared the prize in a contest sponsored by the American Philosophical Society of Philadelphia in 1797 "for the best system of liberal Education and literary instruction, adapted to the genius of the Government of the United States; comprehending also a plan for instituting and conducting public schools in this country, on principles of the most extensive utility." Smith, the young editor of the Jeffersonian *National Intelligencer* of Philadelphia, sought "harmony at home and respect abroad" through a national school system for boys between five and eighteen years of age. He repeatedly expressed his distrust of parents and private, voluntary educational efforts.[13]

Knox, the principal of a boys' academy in Maryland and a Presbyterian minister who publicly embraced liberal religious ideas, denounced ignorance as "the parent and stupid nurse of civil slavery." His linguistic association of "ignorance," "parent," "nurse," and "slavery" perfectly symbolized Knox's suspicion of domestic tutors, private schools, and religious denominational rivalry. Both of these essays, given the stamp of approval of the leading learned society in America, stressed the need for a homogeneous moral education to

[11] Pictured on the cover of this volume

[12] Charles Willson Peale, "Autobiography," 272, in *The Collected Papers of Charles Willson Peale and His Family,* ed. Lillian B. Miller (Millwood, N.Y., 1980), F:IIC; Peale, "To the Citizens of the United States of America," *Dunlap's American Daily Advertiser,* January 13, 1792; Peale to Thomas Jefferson, January 12, 1802, in *Collected Papers,* F:IIA/25A14–B3.

[13] Samuel Harrison Smith, *Remarks on Education: Illustrating the Close Connection between Virtue and Wisdom* (Philadelphia, 1798), reprinted in *Essays on Education in the Early Republic,* 167–223, quotation on 219.

promote the public good rather than the individual development of the scholar.[14]

No republican reformer went farther along the road back to basics than Robert Coram. "No modes of faith, systems of manners, or foreign or dead languages should be taught in those schools," he declared. "As none of them are necessary to obtain a knowledge of the obligations of society, the government is not bound to instruct the citizens in any thing of the kind." Smith joined Coram, Rush, and others in rejecting a classical liberal education (which even Knox defended only above the elementary level) in favor of education for productivity. "Man may indulge himself in sublime reveries, but the world will forever remain uninterested in them," Smith proclaimed. "It is only when he applies the powers of his mind to objects of general use that he becomes their benefactor; until he does this he is neither entitled to their gratitude or applause." Self-indulgence in learning was no more to be tolerated than selfish attachment to parents, religious denomination, or locality; a moral education was to make every citizen of the republic into a paragon of public utility and civic virtue.[15]

Whatever their own backgrounds and training, all of the principal educational thinkers of the early Republic stressed the need for moral education in the new nation. In this sense, educational reformers (and many other Americans) operated within the same mental universe: they shared the basic convictions that (1) low moral character led to poverty, crime, and other social ills; and (2) character was shaped, at least to some extent, by the environment. The first, early-modern view and the second, modern view combined to establish the firm conviction that moral education could solve social as well as individual problems. It could render more wrenching social reforms—and future revolutions—unnecessary.

If the republican reformers envisioned a primary, moral education for all citizens as both one of the "rights of man" and essential for the preservation of republican society and government, their "universal" vision nonetheless had definite limits. In the early Republic, race and gender limited citizenship itself. Thus few intellectuals even considered the question of education for people of color or women. Samuel Harrison Smith judged the issue of "female instruction" to be too controversial for comment. Those who did dare to comment on "female

[14]Samuel Knox, *An Essay on the Best System of Liberal Education, Adapted to the Genius of the Government of the United States* (Baltimore, 1799), reprinted in *Essays on Education in the Early Republic,* 271–372, quotation on 288.

[15]Coram, *Political Inquiries,* 141; Smith, *Remarks on Education,* 198.

education" generally concurred with Noah Webster's conclusion that "in all nations a good education is that which renders the ladies correct in their manners, respectable in their families, and agreeable in society. That education is always wrong which raises a woman above the duties of her station."[16]

Benjamin Rush, who devoted an unusual amount of thought to "female education" as a founding trustee of the Young Ladies' Academy of Philadelphia, did advocate a more comprehensive (albeit "separate and peculiar") education for girls. In an address at the Academy in 1787, Rush argued that, like it or not, American women had to be prepared for "a general intercourse with the world"—for the sake of their husbands and sons. "Female education" must expand beyond the customary reading, writing, and "ornamental accomplishments," to include limited instruction "in the principles of liberty and government" and a thorough inculcation of "the obligations of patriotism"—for the sake of the Republic. "The equal share that every citizen has in the liberty and the possible share he may have in the government of our country," Rush concluded, "make it necessary that our ladies should be qualified to a certain degree, by a peculiar and suitable education, to concur in instructing their sons in principles of liberty and government."[17]

Rush's casual distinction between "citizens" and "ladies" and his expectation that educated women would passively "concur" in men's decisions reveal the boundaries of his educational vision. Women were seen not even as second-class citizens, but merely as the deferential wives and mothers of male citizens. The academy's curriculum of reading, writing, arithmetic, grammar, rhetoric, and geography was not intended to lift women out of domesticity. Certainly, Rush was no advocate of women's liberation, as he hastened to explain.

> I know that the elevation of the female mind, by means of moral, physical, and religious truth, is considered by some men as unfriendly to the domestic character of a woman. But this is the prejudice of little minds and springs from the same spirit which opposes the general diffusion of knowledge among the citizens of

[16]Smith, *Remarks on Education*, 217; Noah Webster, *On the Education of Youth in America* (Boston, 1790), reprinted in *Essays on Education in the Early Republic*, 41–77, quotation on 70.

[17]Benjamin Rush, *Thoughts upon Female Education, Accommodated to the Present State of Society, Manners and Government in the United States of America* (Boston, 1787), reprinted in *Essays on Education in the Early Republic*, 25–40, quotations on 28–29; Rush, *Plan*, 33.

our republics. If men believe that ignorance is favorable to the government of the female sex, they are certainly deceived, for a weak and ignorant woman will always be governed with the greatest difficulty.[18]

As Rush envisioned it, then, the education of American girls was aimed even less at self-realization or self-fulfillment than that of American boys. Indeed, it was doubly removed from such selfish goals, as girls were to be taught to sacrifice their self-interest twice over, first for themselves and then for their sons. This was both necessary and extremely difficult, for according to republican theory, a woman's natural inclination was to disparage civic virtue and to become the "stupid nurse of civil slavery." Even more than their brothers, American girls were to be restrained by their moral education.

Operating on the edge of this intellectual discourse, Judith Sargent Murray's essays on education first challenged and then reinscribed its basic gender assumptions. One of New England's most prolific authors, Murray contributed regular essays to the *Massachusetts Magazine,* a Boston monthly literary magazine published between 1789 and 1796. The second volume of the magazine held her bold manifesto, "On the Equality of the Sexes," published under the pen name Constantia.[19] All but completed in 1779, the essay contests the widespread belief that intellectual differences between men and women are natural or divinely ordained. Rather, Constantia argues, women's lack of education is to blame for their apparent intellectual inferiority: "Are we deficient in reason? we can only reason from what we know, and if an opportunity of acquiring knowledge hath been denied us, the inferiority of our sex cannot fairly be deduced from thence." The social construction of gender roles, not a natural inequality of the sexes, has limited women's intellectual development.[20]

Turning the intellectual discourse on "female education" on its head, Constantia insists that the needs and rights of women should take precedence in shaping their education.

> Should it still be vociferated, "Your domestick employments are sufficient,"—I would calmly ask, is it reasonable, that a candidate for immortality, for the joys of heaven, an intelligent being, who is to spend an eternity in contemplating the works of Deity, should at

[18]Rush, *Female Education,* 79.
[19]Document 10
[20][Judith Sargent Murray], "On the Equality of the Sexes," *Massachusetts Magazine, or, Monthly Museum of Knowledge and Rational Entertainment,* March–April 1790, quotation on 33.

present be so degraded, as to be allowed no other ideas, than those which are suggested by the mechanism of a pudding, or the sewing the seams of a garment?

Allow girls' minds the freedom and nourishment that they need to develop, she argues, and then measure the intellectual equality of women and men. Expanded female education would benefit American society, no doubt, but it should be embraced for the sake of women themselves, in the name of equality and justice.[21]

Although Constantia's radical educational vision continued to resonate through Murray's essays, it was revised and reshaped in later years. In February 1792, Murray launched a new series of essays in the *Massachusetts Magazine* under the pen name Mr. Vigillius or "The Gleaner." Collected in three volumes in 1798 titled *The Gleaner,* these essays continued to disguise Murray's identity, although by this time the source was widely known in Boston. The strategy of assuming a male persona allowed Murray considerable freedom to range over the topics that might interest an eighteenth-century gentleman farmer. Only rarely does "The Gleaner" return explicitly to the subject of education. On those occasions, however, the question of education for girls is moved from the margins to the center of the intellectual discourse.

In an extended fiction called "The Story of Margaretta," "The Gleaner" describes the ideal education of a young girl.[22] The story traces the education of Margaretta Melworth, a young orphan who is adopted by Mr. Vigillius and his wife, Mary. Margaretta's early education falls to Mary, as Mr. Vigillius gives way to "the natural indolence of my temper" and fails to exercise his patriarchal veto.[23] Yet his judgments shape the narrative; the views of Mary and Margaretta are filtered through his perspective, which is itself created by Judith Sargent Murray. "The Story of Margaretta" is thus complicated by a maze of gender identities, as a female author assumes a male voice to narrate a woman's attempts to instruct a young girl. The gender confusion permeating the narrative may have led its more daring readers to ponder the social construction of gender roles in fiction and in life.

Mary's "extensive plan of education" within the domestic sphere introduces Margaretta to religion, needlework, English and French, composition, arithmetic, geography, history, poetry, art, astronomy,

[21] Ibid., 134.
[22] Document 11
[23] [Judith Sargent Murray], *The Gleaner* (Boston, 1798), 3 vols., 1:66–76; reprinted in *The Gleaner* (Schenectady, N.Y., 1992), no. 7, 57–64, quotation on 58.

natural philosophy, music, and domestic economy. It also molds her character by guarding against pride, affectation, and "self conceit," while teaching her to revere herself. Far from being "unfitted for her proper sphere," Margaretta becomes "in every respect the complete housewife," whose puddings are unparalleled. Most important, the narrative suggests, Margaretta's education allows her to resist the flattery of a false suitor, to marry an honorable neighbor, and to become "a pleasing and instructive companion" to her husband and family.[24]

If the character of Margaretta embodies Murray's mature educational philosophy, she certainly does not realize Constantia's radical vision of 1779. Margaretta's education resembles that offered in Benjamin Rush's Young Ladies' Academy and is justified in similar terms: it prepares her for marriage and a productive domestic life. In fact, Margaretta never leaves the domestic sphere, although that sphere expands in importance as her intellectual abilities develop. Shadowing her adoptive mother, she might one day use her knowledge to shape the mind and character of her own daughter. In her circumscribed but vital role as a republican wife and mother, she can serve the Republic without enjoying the rights of citizenship. Although Murray's earlier theme of female equality and self-fulfillment never quite disappears, it is certainly muted in this narrative. "The Gleaner's" retreat from Constantia's radical stance underscores the persuasive power and influence of the republican educational vision in the 1780s and 1790s.

Margaretta's education is conducted primarily through reading, conversation, writing, and the subtle indoctrination characteristic of maternal nurture. After all, her teacher is also her mother. In another essay, "The Gleaner" protests against the more severe teaching methods commonly used in American schools and academies. "The austere man can never be successful," the essay insists; "he will banish smiles from the face of that season which is made for joy; and if the student is not uncommonly endowed by nature, he will create in him an aversion to his book." None of the republican educational reformers shared "The Gleaner's" opposition to traditional modes of instruction or strict discipline. They distrusted maternal nurture far more than the rod.[25]

Indeed, the republican reformers' theoretical emphasis on deliberate restraint and moral inculcation was reflected strongly in the pedagogical methods they favored. Samuel Harrison Smith would have his

[24]Ibid., 58, 63, 61, 59.
[25]Ibid., no. 35, 286–92, quotation on 289.

boys memorize and recite "select pieces, inculcating moral duties," and the U.S. Constitution from age five to eighteen. Samuel Knox wished to have his primary pupils "carefully commit to memory the rules in the various branches" of English grammar and mathematics "and rehearse these rules once a week." Later these basic exercises could be supplemented with "a well-digested, concise moral catechism."[26]

Entirely consonant with his general view of education to meet a predetermined external goal was Benjamin Rush's insistence on a strong figure of authority in the school. "The government of schools . . . should be *arbitrary*," Rush held. "By this mode of education, we prepare our youth for the subordination of laws and thereby qualify them for becoming good citizens of the republic. I am satisfied that the most useful citizens have been formed from those youth who have never known or felt their own wills till they were one and twenty years of age. . . . "[27]

Rush, whose medical training and general enlightenment in Edinburgh had led him to embrace the Scottish doctrine of an innate moral sense in man, nonetheless continued to fear man's natural "will" or passions. This unlikely philosophical combination of Scottish optimism and traditional republican (and Christian) pessimism about human nature apparently impelled him to advocate simultaneously the full development of man's moral faculty and the vigorous repression of man's natural will. Lifelong habits could be built on either part of human nature, for good or ill.

Other, more conservative American republicans pushed the pedagogical emphasis on restraint still further. Noah Webster, the tireless "Schoolmaster to America" from Connecticut, carefully constructed his elementary grammars and readers to include both the traditional rules and definitions to be memorized, and new didactic moral tales and patriotic names and deeds suitable for "lisping" at an early age. His ubiquitous *American Spelling Book* of 1788, which was selling more than two hundred thousand copies a year by 1807, featured both "A Moral Catechism" and "A Federal Catechism."[28] From these, the youngest children could memorize and recite the advantages of virtuous behavior and republican government. Moreover, when gentle suasion failed, Webster was ready to resort to the discipline of "the rod"

---

[26]Smith, *Remarks on Education*, 211; Knox, *Essay on the Best System*, 330, 332.
[27]Rush, *Plan*, 24.
[28]Document 12

to maintain "strict subordination" and the "absolute command" of the schoolmaster. Even more than Smith, Knox, or Rush, Webster relied on traditional pedagogical methods and a strong external moral authority to create republican citizens in America.[29]

Overall, then, the intellectuals of the early Republic shared a broad vision of education, which included intellectual development but focused primarily on moral education for citizenship. Above all, they favored schooling that would promote the cultivation of civic virtue, or the willingness to sacrifice natural self-interest and familial or local attachments for the public good. This predetermined goal led the republican educational reformers to advocate a strong (and almost invariably male) figure of authority in schools and to license various forms of discipline to maintain proper "subordination." Their well-defined external goal also led the intellectual elite to embrace traditional pedagogical methods, including memorization and recitation of various rules, definitions, and republican "catechisms."

Displaying at different times a belief in a Lockean "blank slate," a faith in an innate moral sense, and a fear of natural will or passions, these American intellectuals sought to allow for all three possibilities in their educational theory. The result was a complex moral education to be imposed on all future citizens by a strong external authority, the repression of errant passions, and the inculcation of republican principles. Together, these rather traditional methods were to fortify the moral faculty, restrain the natural passions, and form the lifelong habits of "republican machines."

[29]Noah Webster, *The American Spelling Book* (Boston, 1798), 154–56; Webster, *Education of Youth*, 57–58.

4

Narrating Nationhood

In their most self-conscious moments, American intellectuals stared into "the mirror which reflects the true image of a nation" and began to compose their collective autobiography—the history of the American Republic. Their mission was exalted, the historians knew, for they offered their countrymen self-knowledge, moral instruction, and pleasure. Because "the actions and affairs of men are subject to as regular and uniform laws, as other events" in nature, the study of history is "the most important of all our philosophical speculations" and a "source of the sublimest moral improvement," one New England minister/historian declared. According to another, history "teaches human nature, politics and morals, forms the head and heart for usefulness, and is an important part of the instruction and literature of states and nations." Indeed, it was difficult to imagine a nation without a history of its own. As the editor of a leading Philadelphia literary magazine concluded, "every genuine patriot" must "earnestly long" for an authentic American history.[1]

Thus the intellectuals of the early Republic approached historical writing with a double vision. Although they believed that history must reflect "the true image of a nation," they also understood that historians created and projected a national identity for domestic and foreign consumption. History must be truthful, but it also must be instructive, able to help those inside and outside the nation to understand its identity and promise. The realization that different interpretations of the past could emerge from different quarters was inescapable for those American intellectuals who read British and Loyalist accounts of the American Revolution. The fiercely anti American *Political Annals of the Present United Colonies,* published in London in 1780, illustrated

[1]*American Museum* (Philadelphia, 1787–92), 11:43–46, 10:145–46; Samuel Williams, *The Natural and Civil History of Vermont* (Walpole, N.H., 1794), xi; Benjamin Trumbull, *A Complete History of Connecticut, Civil and Ecclesiastical* (Hartford, 1797), 1:iii.

the danger of allowing outsiders—geographical or temperamental—
to write American history. If they wished to preserve their image
unsullied by European incursions, Americans would have to develop
their own narratives of nationhood.

Most of the histories written in the early Republic were local or
regional, and all were written by amateurs. In an age and place without professional historians, history was written as an act of love, and
that love was usually directed toward a locality or state. Following in
a long colonial tradition begun by the Pilgrim leader William Bradford, the ministers, statesmen, physicians, and women of the early
Republic wrote histories to record the trials and virtues of their ancestors, friends, and neighbors. As the Reverend Jeremy Belknap noted
of his *History of New Hampshire,* "The particular incidents of these
wars, may be tedious to strangers, but will be read with avidity by
the posterity of those, whose misfortunes and bravery were so conspicuous." In eighteenth-century style, the historians' chronicles of
events were punctuated by philosophical reflections on the moral
lessons they contained. Between 1775 and 1800, richly detailed historical narratives of New Hampshire, Vermont, Connecticut, South Carolina, and other states were written and published primarily for local
audiences.[2]

As appreciative as American intellectuals were of these efforts to
construct a usable past, many believed that local history was too limited to satisfy the new nation's needs. The intellectuals of the early
Republic greeted each state history with admiration and calls to
broaden the geographical scope. They worried that memories would
fade, the principal actors would die, and the grand philosophical
lessons to be drawn from the American Revolution would soon be lost.
They had no doubt that a nation must have a national history. As political and social harmony seemed ever harder to achieve, the demand
for someone to narrate nationhood grew stronger. If historians could
conceive of America as a nation and give it a national history, perhaps
a sense of national identity and unity would follow.

Most of the amateur historians shrank from the daunting prospect
of writing a national history before the various state histories were
compiled or documents collected. Despite specific requests from "the
General Association of the State" of Connecticut and others for "a gen-

<hr>

[2]Jeremy Belknap, *The History of New Hampshire* (Boston, 1784, 1791, 1792), 3 vols.,
1:iii.

eral history of the United States of America," the Reverend Benjamin Trumbull confined himself to regional history. Even so, the first volume of his *Complete History of Connecticut* was twenty-three years in the making. Trumbull considered it "not very consistent with that respectful and generous treatment which he owed more particularly to his own state, to publish a large history of the United States, while he neglected theirs." The historian of Vermont, the Reverend Samuel Williams, also found the local approach most appealing. "To represent the state of things in America in a proper light, particular accounts of each part of the federal union seem to be necessary," he explained. Reflecting patterns of allegiance as well as knowledge, most historians of the early Republic resisted the call to broaden the scope of their historical narratives.[3]

Yet a few did rise to the oft-repeated challenge. The acknowledged leader in the effort to narrate nationhood was David Ramsay of South Carolina. Known to his contemporaries as a statesman, physician, and the "American Tacitus," Ramsay wrote two histories of South Carolina, a history of the American Revolution, and even a twelve-volume *Universal History Americanized*. He devoted much of his life to the creation of a unified American culture, and he recognized that history could play a special role in constructing a sense of national identity and destiny for his contemporaries and posterity alike.

Born and raised in Lancaster County, Pennsylvania, Ramsay graduated from the College of New Jersey in 1765 and, after teaching briefly in Maryland, studied medicine at the College of Philadelphia. There he befriended Benjamin Rush, who warmly recommended Ramsay for a medical position in Charleston, South Carolina. Rush himself had recently refused the post, for its locus was repugnant to his antislavery principles. Ramsay moved to Charleston in 1774.

Almost from the moment he arrived, Ramsay found himself caught up in politics. In February 1776—the year in which he was elected to the South Carolina assembly—he wrote to Rush to applaud the "venerable" author of *Common Sense* (Thomas Paine) and to express his own radical political views.

> I trust that America will eventually prove the asylum for liberty learning religion & et & that the civil & religious rights of mankind will be most effectually guarded by our democratic legislatures. A glorious exploit to redeem the eighth part of the habitable globe

[3]Trumbull, *History of Connecticut*, 1:v–vi; Williams, *History of Vermont*, viii.

from tyranny oppression & suppression. A case to fight for—to bleed for—to die for.[4]

The same spirit inspired Ramsay's first history, composed during his continuous wartime service in the South Carolina legislature, his exile to St. Augustine, Florida, with other prominent South Carolinians in August 1780, and his service as a delegate to the Confederation Congress in 1782 and 1785. Finally published in 1785, his *History of the Revolution of South Carolina* sought to preserve the minute details of these exhilarating events from being lost in the tumultuous current events of the 1780s. Its ardent patriotism is unmistakable. Aimed abroad as well as at a local audience, the book openly proclaims the justice of South Carolina's patriot cause, the virtue of its rebels, and the vice of its "royalists." It was yet another barrage in defense of the Revolution.[5]

Encouraged by Thomas Jefferson and Philip Freneau, among others, to "extend his plan"—and seeking to sustain the excitement of the Revolutionary years during long, dull hours in the Congress and the South Carolina legislature between 1784 and 1789—Ramsay collected material for another history.[6] In many ways, his pioneering *History of the American Revolution* of 1789 departs from his earlier history.[7] Tripping a bit along the way, it begins to narrate nationhood.

Hoping to reach a national audience, Ramsay expanded both his geographical and chronological scope. This presented new perils as well as opportunities for the historian. Beginning with a long chapter on "the Settlement of the English Colonies, and of the political Condition of their Inhabitants," he quickly discovered the difficulty of painting them all with a single brush. Try as he might, he could find no single pattern of settlement or virtuous principle animating the whole. Indeed, the greater expanse of his subject and his emotional distance from most parts of it led Ramsay to notice contrasts and shades of

[4]David Ramsay to Benjamin Rush, Charlestown [Charleston, S.C.], February 14, 1776, in *David Ramsay, Selections from his Writings,* ed. Robert L. Brunhouse, Transactions of the American Philosophical Society, n.s., 55, pt. 4 (Philadelphia, 1965), 53 [hereafter *Selections*].

[5]David Ramsay, *History of the Revolution of South Carolina* (Trenton, N.J., 1785), 2 vols., I:v.

[6]Thomas Jefferson to David Ramsay, Paris, August 31, 1785, in *Selections,* 92; Philip Freneau, "On the Legislature of Great-Britain Prohibiting the Sale, in London, of Dr. David Ramsay's History of the Revolutionary War in South Carolina," in *American Museum* (Philadelphia, 1787), 1, no. 2.

[7]Document 13

virtue within the colonies. A comparative stance could lead to an unexpected critique of colonial actions:

> Mankind then beheld a new scene on the theatre of English America. They saw in Massachusetts the Puritans persecuting various sects, and the church of England in Virginia, actuated by the same spirit, harassing those who dissented from the established religion, while the Roman Catholics of Maryland tolerated and protected the professors of all denominations.

Perhaps unintentionally, the text reveals the follies, weaknesses, and cultural diversity of the English colonies in America.[8]

At times, Ramsay is clearly troubled by his own comparisons. He could not help noticing that "though the Southern Provinces possessed the most fruitful soil and the mildest climate, yet they were far inferior to their neighbors in strength, population, industry, and aggregate wealth." South of Maryland, he could find no equivalent to the general "industry and morality" of early New England or the "Quaker simplicity, industry, and frugality" of early Pennsylvania. Locating the source of this pervasive cultural difference in the colonies' labor systems, Ramsay indicts domestic slavery less for its injustice to the enslaved than for its "baneful consequences" for the masters. "Idleness is the parent of every vice, while labour of all kinds favours and facilitates the practice of virtue," he lectures his readers. "Unhappy is that country, where necessity compels the use of slaves, and unhappy are the people, where the original decree of heaven, 'that man should eat his bread in the sweat of his face' is by any means whatever generally eluded."[9]

Pausing to reflect on the meaning of his story so far, Ramsay acknowledges its fundamental ambiguity. On the one hand, Europe had benefited enormously from American wealth and industry. The "prodigious extension of commerce, manufactures, and navigation, and the influence of the whole on manners and arts," seem undeniable blessings. However, the text immediately adds, "when we view the injustice done the natives, the extirpation of many of their numerous nations, whose names are no more heard" or the "slavery of the Africans, to which America has furnished the temptation," we begin to suspect "that the evil has outweighed the good." By the admission of

[8]David Ramsay, *The History of the American Revolution* (Philadelphia, 1789), 2 vols., 1:11.

[9]Ibid., 1:25, 20, 21, 23–24.

its author, the narrative of American nationhood was off to a shaky start.[10]

Yet all was not lost. These interpretative difficulties did not deter Ramsay from finding the roots of American identity in the colonial past. Transcending the "tedious" and "unprofitable" details of colonial political and constitutional history, the narrative introduces a new mythical figure, the "American Colonist." Distance from the source of power in England, the vastness of the American landscape, a general social equality in the colonies, and the Protestant religion all produced in this Colonist "a warm love for liberty, a high sense of the rights of human nature, and a predilection for independence." Gendered male throughout the text, the Colonist was, "or easily might be," a hardworking small farmer and independent freeholder. He looked to heaven rather than to kings for protection. Whatever his ethnicity, sect, or region, he shared with his fellow colonists a set of political ideas that formed the basis of American identity.

> The political creed of an American Colonist was short but substantial. He believed that God made all mankind originally equal: that he endowed them with the rights of life, property, and as much liberty as was consistent with the rights of others. That he had bestowed on his vast family of the human race, the earth for their support, and that all government was a political institution between men naturally equal, not for the aggrandizement of one, or a few, but for the general happiness of the whole community.

In short, the American Colonist was a strong, independent, white, male republican, unmarked by that debilitating slave system already gripping the southern colonies.[11]

The creation of this mythical figure restructures the narrative. A new clarity and coherence emerges, as the American Colonist pushes all other narrative lines to the margins. No longer mired in the ambiguity and diversity of the earlier pages, the pace of the story quickens. Indeed, as it traces the origins and course of the Revolutionary conflict, the narrative acquires the relentless quality of manifest destiny. By 1763, the argument goes, the English colonies in America "had advanced nearly to the magnitude of a nation," although few seemed to notice. Misguided British imperial policy thereafter led to rapid colonial alienation and independence. The text notes without surprise

---

[10]Ibid., 1:13–14.
[11]Ibid., 1:17, 26, 32, 31. See also Arthur H. Shaffer, *To Be an American: David Ramsay and the Making of the American Consciousness* (Columbia, S.C., 1991), 105–27.

that "the seeds of discord were soon planted, and speedily grew up to the rending of the empire." After all, "combustible materials" had long been "collecting in the new world," wanting only "a spark to kindle the whole" from the old. When that spark flew, an inchoate American identity was already in place, waiting for the shared experiences of the Revolution to realize and fulfill it.[12]

Ironically, the inexorability of this narrative line permits the historian considerable flexibility in the presentation of peripheral issues. Because the creation of an American national identity is prefigured in this interpretation, other controversial topics can be approached with subtlety and balance. Thus British imperial policy is viewed as unwise and inconsistent, but not corrupt or tyrannical. Indeed, the same principle of human nature is invoked to explain British and colonial actions on the eve of the Revolution: "The love of power, and of property, on the one side of the Atlantic, were opposed by the same powerful passions on the other." Similarly, the text refuses to vilify all Loyalists, acknowledging that some acted on religious or political conviction. If these interpretations are unusually nuanced for the period, it may be because they could not alter the predetermined outcome of the history's central narrative.[13]

The last two chapters of *The History of the American Revolution* recapitulate the general narrative pattern and underscore the central plot. Once again, the Revolution (and its logical culmination, the federal Constitution of 1787) allowed the colonists to recognize and celebrate their shared identity. Colonial cultural differences and divisions are briefly acknowledged, only to be brushed aside once more as peripheral to the plot.

In a retrospective penultimate chapter, Ramsay offers reflections on "the State of parties: the advantages and disadvantages of the Revolution: its influence on the minds and morals of the Citizens." Beginning with an analysis of ethnic, religious, and social divisions within Revolutionary America, he frankly admits that unity of purpose "could not reasonably be expected." Yet the cool, rational analysis gradually gives way to a characterization of Whigs as young, ardent, ambitious, and enterprising, and Tories as phlegmatic, timid, self-interested, and indecisive. The American Colonist had clearly become a revolutionary Whig.[14]

---

[12]Ramsay, *History of the American Revolution,* 1:34, 42.
[13]Ibid., 1:53, 125.
[14]Ibid., 2:310, 314.

Shifting the narrative focus to these active patriots, the chapter then explains how such ordinary "self-made, industrious men" became statesmen and leaders in a period of crisis. In the Continental army and Congress, their provincial perspectives were enlarged, and they began to recognize and articulate their common values. Their transformation from colonists to Whigs to Americans during the Revolution lies at the heart of Ramsay's story, even in a chapter that begins with partisanship. Celebrating the "establishment of a nation out of discordant materials," the narrative clearly privileges unity over diversity and consensus over conflict.[15]

The chapter ends on a cautionary note: American unity was still quite fragile in 1783. The contagion of liberty had infected "the multitude," and few traditional authorities remained strong enough to fend off moral decay. Worrying that "a long time, and much prudence, will be necessary to reproduce a spirit of union and that reverence for government, without which society is a rope of sand," the narrator suggests that the Revolution might have no end.[16]

This apprehension subsides only in the final chapter of the history, as Americans "unanimously turned their eyes" to George Washington to preside over their reconstituted federal government in 1789. Union is as much the hero as Washington in this grand climax to the Revolution and the book. As the historian urges Americans to practice "the whole lovely train of republican virtues," to seek "universal justice," to cultivate "union between the East and South, the Atlantic and the Mississippi," and to aspire to "national greatness," America's first narrative of nationhood draws to a triumphal close.[17]

Ramsay's principal rival as historian of the American Revolution was also well positioned to fashion a narrative of nationhood. One of the leading female intellectuals and most prolific authors of the early Republic, Mercy Otis Warren was a member of New England's social and political elite. Her marriage to James Warren in 1754 joined two of the most prominent and powerful families in Massachusetts. The two families' dedication to public service continued during the Revolution: Mercy Otis Warren's brother, James Otis, was one of the colony's leading legislators and orators; her husband helped found the Massachusetts Committee of Correspondence and served as president of

[15]Ibid., 2:316.
[16]Ibid., 2:324, 323.
[17]Ibid., 2:344, 354–56.

the Massachusetts Provincial Congress; and his cousin, General Joseph Warren, fell at the Battle of Bunker Hill.

Barred from direct political participation, Mercy Otis Warren poured her own considerable energy and intellect into her writing. She regarded historical writing as a form of public service that women could perform as well as men. As early as 1775, her close friend Abigail Adams expressed the hope that Warren's "Historick page will increase to a volume." Abigail's husband, John, echoed that wish in 1787: "Your Annals, or History, I hope you will continue, for there are few Persons possessed of more Facts, or who can record them in a more agreeable manner." By 1791, Warren had finished all but the last chapter of her three-volume history.[18]

Warren's *History of the Rise, Progress and Termination of the American Revolution,* finally published in 1805, stands with Ramsay's as the most important contemporary history of the Revolution.[19] It also constructs a narrative of nationhood, though of rather different materials. Like Ramsay, Warren infuses her history with philosophical and moral observations and lessons in republicanism. Unlike Ramsay, however, she paints a stark, dramatic historical portrait, in which the American patriots are moved by virtue and a love of liberty to oppose an ungrateful, dissipated, and corrupt British nation and its minions. The historical construction of larger-than-life, legendary Revolutionary figures begins here, as Warren weaves the classical republican oppositions of virtue and vice, liberty and power, into the web of her narrative of nationhood.

As staunch an Antifederalist as Ramsay was a Federalist, Warren appears untroubled by the spread of libertarian and egalitarian sentiments during the Revolution. American unity is far less important to her, and diversity less problematic. Thus she has no need for the mythological figure of the American Colonist to shape her narrative. As long as particular Revolutionary characters with local loyalties, universal visions, and idiosyncratic personalities are strongly drawn, they serve just as well to forward her conception of a nation of liberty-loving individuals.

Warren unwaveringly places liberty and its defenders at the core of her history. This is not necessarily a parochial gesture. She finds

---

[18]Abigail Adams to Mercy Otis Warren, November 1775, *Warren-Adams Letters,* Massachusetts Historical Society, *Collections,* vols. 72–73 (1917, 1925), 1:179; John Adams to Mercy Otis Warren, December 25, 1787, ibid., 2:301.

[19]Document 14

liberty and virtue wherever American patriots struggle against the British foe, and her narrative follows the Revolutionary political and military conflict from the Stamp Act through the adoption of the federal Constitution of 1787. Yet her broad geographical and chronological scope does not keep Warren from giving disproportionate space and admiration to those New Englanders (including her friends, neighbors, and family) who, the narrative forcefully suggests, courageously lead a nation in the path of virtue. Liberty's heroes often speak with a Massachusetts accent in the *History of the Rise, Progress and Termination of the American Revolution*. Establishing a long tradition in early American historiography, Warren demonstrates here that a narrative of nationhood might be constructed out of local materials, particularly if that locality is New England.

Surely the master historical mythmaker of the period—and perhaps of all time—was Mason Locke Weems. Born in Maryland in 1759, Weems spent most of the Revolutionary War years in England studying medicine and preparing for the ministry. He was ordained a priest in the Church of England in 1784 and returned home to preach in rural Maryland and Virginia. Although he never served as rector at Mount Vernon (as he would claim in later years), he was well positioned to collect stories about George Washington and his family. By 1794, the country parson had also become an itinerant bookseller for the Philadelphia publisher Mathew Carey. Traveling repeatedly between New York and Georgia, Weems learned a great deal about his customers' tastes and desires. He found it impossible to keep enough novels in stock and ever more difficult to sell sober religious tracts. He was not one to ignore such lessons.

Weems saw in Washington's death in December 1799 an unparalleled commercial opportunity as well as a chance to offer Americans instruction in public and private virtue and the need for national unity. He seized the opportunity with enthusiasm. By mid-January 1800, he explained his plan to Carey:

> I've something to whisper in your lug. Washington, you know is gone! Millions are gaping to read something about him. I am very nearly prim'd and cock'd for 'em. 6 months ago I set myself to collect anecdotes of him. My plan! I give his history, sufficiently minute—I accompany him from his start, thro the French & Indian & British or Revolutionary wars, to the Presidents chair, to the throne in the hearts of 5,000,000 of People. I then go on to show that his unparrelled [*sic*] rise & elevation were due to his Great Virtues.

When Carey failed to respond by the end of the month, Weems "resolv'd to strike off a few on my own acc't."[20]

Weems's *Life of George Washington; with Curious Anecdotes, Equally Honourable to Himself and Exemplary to his Young Countrymen* of 1800 succeeded beyond even his wildest dreams.[21] A runaway bestseller that went through nine editions by 1809, the slim volume combines patriotism, sentiment, and religiosity to create both an exemplary life (on the order of an American saint's) and an enduring image of the "Father of His Country." It also constructs a narrative of nationhood. In offering his readers a national hero whose paternal authority and care for all Americans placed him high above sectional loyalties or partisan squabbling, Weems forged a sense of national identity and unity for his own and succeeding generations.

Although most of the biography traces Washington's illustrious public career, the first four chapters focus on the formation of his character in childhood and youth. They engaged readers from the start and were the first part of the biography to be expanded. As Weems explains, the emphasis in these chapters on Washington's "private virtues" allows every American to identify with the future hero, "because in these every youth may become a Washington."[22] The early chapters also help bridge the gap between the period's classical republicanism and evangelical Protestantism. As different as these systems of belief were, both required individuals to conquer selfishness and private vices before performing on a public stage. The first chapters could thus invite readers of diverse cultures to enter into the narrative of America's national hero.

The opening chapters of the *Life of George Washington* achieved one thing more. When the celebrated fiction of young George's encounter with his father's cherry tree was introduced in the fifth edition of 1806, the emerging narrative of nationhood took a new turn. Weems presents this lesson in integrity in a form borrowed from popular sentimental novels of the day. This experiment in cultural fusion was audacious, for such novels were profoundly distrusted by intellectuals of the early Republic. To instruct America's youths in integrity through fiction took Weems's confidence and shrewdness. Once

[20] Mason Locke Weems to Mathew Carey, "Jan. 12 or 13," 1800, and February 2, 1800, in *Mason Locke Weems: His Works and Ways*, 3 vols., ed. Emily E. Ford Skeel (Norwood, Mass., 1929), 1:8–9.

[21] Document 15

[22] Mason Locke Weems, *The Life of George Washington; with Curious Anecdotes, Equally Honourable to Himself and Exemplary to his Young Countrymen*, 9th ed. (Philadelphia, 1809), 7.

again, his instincts proved sound. His amusing little anecdote not only drew readers into his narrative of nationhood; it also created one of the most unshakable myths of American culture.

Some of the artists of the early Republic also constructed narratives of nationhood in their work. In historical paintings and portraits of Revolutionary leaders, they created a series of visual images that conveyed a sense of national identity. Like Ramsay, Warren, and Weems, these artists stressed the common political values, shared Revolutionary experiences, and national heroes that might bind a diverse nation together. Moreover, because visual representations (especially when on public display or engraved and reproduced) could reach a wider audience than the written word, the artists' narratives of nationality commanded significant attention.

The most prolific and respected painter of the American past was John Trumbull. A native of Connecticut and graduate of Harvard College, Trumbull joined the Continental army in Boston in 1775. He served as an aide-de-camp to General George Washington and as adjutant to General Horatio Gates. His father, Jonathan Trumbull, was governor of Connecticut before and during the Revolution. In 1777, John Trumbull left the army to paint, first at home, then in Boston, and finally in Paris and London, where he studied with the American expatriate Benjamin West after the war. When he realized that West, a court painter to King George III and president of the Royal Academy in London, could not credibly interpret recent history on canvas for Americans, Trumbull decided to try.

In a remarkable series of historical paintings completed between 1786 and 1796, Trumbull captured the drama and dignity of the Revolution. The first of these visual representations of American "national history," *The Death of General Warren at the Battle of Bunker's Hill,* was painted in London between 1784 and 1786.[23] Both realistic and highly dramatic, it drew on the artistic innovations of Benjamin West and John Singleton Copley (another American expatriate living in London), whose history paintings incorporate accurate background details, portraits, and modern dress. The painting also drew on Trumbull's own experience: he had watched the battle from a distance in 1775.

Like Warren's history, Trumbull's painting emphasizes the heroism of the citizen-soldiers who valiantly defended their principles and country against Britain's professional army. This narrative point is made visually both by the composition of the painting (which places

[23]Document 16

the outnumbered patriots in a defensive posture at the front and center) and the contrasting costumes of the two groups of soldiers (the British in full military dress, the patriots in civilian clothing). The painting's brilliant colors, lively brushwork, and diagonals of banners and smoke all work together to convey the drama of the battle. The small size of the canvas (twenty-five by thirty-four inches), designed to accommodate engravers and to encourage print reproductions, indicates Trumbull's desire to narrate nationhood for the widest possible audience.

A year later, Trumbull completed *The Declaration of Independence, Philadelphia, 4 July 1776.*[24] The painting includes a life portrait of John Adams, who stands in the central group with the other principal founders. While *The Death of General Warren* is all swirling motion and color, *The Declaration of Independence*'s linear, horizontal composition and sober, formally attired, stationary figures convey a sense of solidity and strength. In this visual narrative, the dignity and inevitability of the birth of a nation replace the drama and contingency of the earlier battle scene. American independence, the painting suggests, depends as much on ideas, books, and papers as on swords, guns, and smoke. Engraved and reproduced many times, Trumbull's *Declaration of Independence* was also copied by the artist himself onto the walls of the central rotunda of the U.S. Capitol in 1817. His representation thus became part of the official narrative of American identity.

In the developing iconography of nationhood, portraits of the founders took pride of place. And judging from the number of portraits commissioned, painted, and copied, George Washington's was the face of the nation. Washington appeared in a seemingly endless variety of poses and settings: more than eight hundred prints from engravings of his portraits were made in the early years of the republic.

The artistic apotheosis of Washington began during the war itself. Charles Willson Peale, an officer in the Pennsylvania militia, political activist in Revolutionary Philadelphia, and friend of many of the founders, painted several life portraits of Washington in the 1770s. The most impressive of these is *George Washington at the Battle of Princeton,* an official portrait commissioned by the Supreme Executive Council of Pennsylvania in January 1779.[25] Copied two dozen times by the artist himself, the painting enjoyed great popularity both as a commemoration of the American military victories at Trenton and

[24]Document 17
[25]Document 18

Princeton (in which Peale had participated) and as an icon of Washington as a military leader.

The painting narrates American nationhood in both large and small details. Framing the picture at the right, an American battle flag waves triumphantly over folded Hessian standards captured at Trenton. In another symbol of American power, Continental soldiers lead away their British prisoners in the background. Dominating the painting, a monumental full-length portrait of Washington projects an air of confidence and ease through its informal pose. With his hat off and hand resting lightly on a cannon, America's military commander seems both strong and comfortable in his position of leadership. His nobility of character is offset by his apparent egalitarianism, as his eyes directly meet the viewer's gaze. Peale's meticulous attention to detail, evident in his removal of a blue ribbon on Washington's chest to accord with a change in the military uniform code in June 1780, also supports his narrative of nationhood. America not only had valiant citizens who rose to defend their homes, the painting suggests, it also had a victorious uniformed army led by a confident and implicitly democratic citizen-commander.

Two decades later, another monumental full-length portrait of Washington reshaped his image to meet the country's developing vision of its national hero. Painted by Gilbert Stuart, regarded as the finest American portraitist of the day, this image acquired special significance, even official status, in the emerging iconography of nationhood. Although he hated to pose for portraits, Washington sat for Stuart, who returned home from England expressly to paint him. After completing the portrait and two busts in 1795–97, Stuart reproduced them more than one hundred times over the next thirty years. One of the first copies of the full-length portrait was purchased by the U.S. government in 1800 to display in the White House. Considered a national treasure, it was the only painting saved by Dolley Madison as she fled before the British forces who set fire to the White House in August 1814.

Even more than the famous busts, Stuart's full-length portrait, titled simply *George Washington,* created the authoritative image of America's first president.[26] Dressed in civilian clothing and surrounded by symbols of his office, Washington grasps a sheathed sword; he is commander in chief, but no longer a military figure. His arm is suspended over a table laden with pen, papers, and two books, *The Federalist*

[26]Document 19

Papers and *The Journal of Congress.* A folio volume of the *Constitution and Laws of the United States* and a history of the American Revolution support the table leg, which is adorned with two American eagles. Behind the imposing central figure, the seat of authority and a column of order structure the composition. The power of ideas and law in a well-constituted state surrounds the president and anchors this narrative of nationhood.

Washington himself is a dignified, commanding presence in the painting. The ease and informality of Peale's Washington are gone, replaced by an expression of unbending resolve and self-control. As Stuart explained his conception, "All his features were indicative of the strongest passions; yet like Socrates his judgment and self-command made him appear a man of different cast in the eyes of the world." Conscious that the eyes of the world are upon him and the nation he embodies, the president stares off into the distance, perhaps toward posterity. Now older, more austere and idealized than in Peale's portrait, Stuart's Washington has become the "Father of His Country."[27]

[27]Gilbert Stuart, quoted in William Kloss et al., eds., *Art in the White House: A Nation's Pride* (Washington, D.C., 1992), 68.

5

Contesting Popular Culture

Even as the poets, educators, historians, and artists labored to create an American culture, American readers and viewers began to express their preference for other cultural forms. To the intellectuals' dismay, the American public seemed to crave those sentimental novels and plays that also enjoyed great popularity in England in the late eighteenth century. That almost all of the novels and plays available in the 1780s were English apparently didn't bother their growing American audience at all. But when American imitators appeared in the late 1780s and 1790s, Americans flocked to their local theaters, booksellers, and lending libraries. An American popular culture, based on sentimental English models, was born.

The intellectual elite viewed the rise of this popular culture as a crisis. Even those intellectuals whose republican visions diverged most widely agreed that the emerging culture—which they believed to be largely inspired, consumed, and even produced by women—was one of the principal dangers facing the nation. Ministers, lawyers, physicians, statesmen, and college presidents of the early Republic all voiced their opposition to it. John Witherspoon, the influential president of the College of New Jersey and a teacher of James Madison, Philip Freneau, and Hugh Henry Brackenridge, summed up the views of many when he declared that "romances and fabulous narratives are a species of composition, from which the world hath received as little benefit, and as much hurt as any . . . excepting plays themselves."[1]

The intellectuals' suspicion of novels and plays was also inscribed in law. In 1775, the Continental Congress passed a resolution to "discountenance and discourage every species of extravagance and dissipation, especially all horseracing, and all other kinds of gaming,

---

[1] *United States Magazine* (Newark, 1794), 1:245. Also weighing in with authoritative pronouncements against sentimental fiction were John Adams, Thomas Jefferson, Timothy Dwight, Jonathan Trumbull, Hugh Henry Brackenridge, Noah Webster, Royall Tyler, and Benjamin Rush, among others.

cockfighting, exhibitions of shews [sic], plays, and other expensive diversions and entertainments." Even after the war, state assemblies and local governments often debated and sometimes banned the opening of theaters. The Common Council of New York City closed the John Street Theater, a "fruitful source of dissipation, immorality, and vice," soon after its opening in 1785. Yet America's intellectual and political leaders could not hold back the tide forever. Between 1785 and 1812, restrictive legislation was repealed and new theaters constructed in New York, Philadelphia, Charleston, Baltimore, Washington, and even Boston, where a ban of more than forty years was lifted in 1793.[2]

Modern readers of the extremely didactic novels and plays of the early Republic may well be mystified by this widespread and deep concern among American intellectuals. Even sentimental novelists of the period loaded their volumes with far more blatant didacticism and moral instruction than their European counterparts. Characters of various novels themselves commented on this phenomenon and complained that American novels were hardly more entertaining than sermons. Indeed, the most prolific and popular sentimental novelist of the early Republic fulfilled her felt duty to instruct "the female part of the rising generation" in morality both as a novelist and as a schoolteacher. Early American plays, too, were prefaced by serious declarations of moral duty and service to the Republic. Neither seemed likely to provoke strong antipathy or fear.[3]

Yet the intellectual elite believed that America's moral health was at risk. All of the moralistic prefaces and prologues in the world, they insisted, could not counteract the dangerous tendencies of fiction to subvert moral virtue. As one widely reprinted article of the 1790s warned, "It is in that school the poor deluded female imbibes erroneous principles, and from thence pursues a flagrantly vicious line of conduct." The article's argument and title, "Novel Reading, a Cause of Female Depravity," suggested that fiction directly threatened the virtue of the simple, innocent young women who were its principal consumers. And that threat was perceived to be a potent one, because "principles" were thought to lead inexorably to "conduct." As another widely printed article of the 1790s explained, "Novels not only pollute

---

[2]*Journals of the Continental Congress, 1774–1789,* ed. Worthington Chancey Ford (Washington, D.C., 1904), 1:78; William Dunlap, *A History of the American Theatre* (New York, 1832), 59.

[3]Hannah Foster, *The Boarding School* (Boston, 1798), 156–57; Susanna Haswell Rowson, *Reuben and Rachel; or, Tales of Old Times* (Boston, 1798), 1:iii.

the imaginations of young women, but likewise give them false ideas of life, which too often make them act improperly; owing to the romantic turn of thinking they imbibe from their favourite studies." In keeping with their optimistic faith that true republican culture could foster virtuous behavior, these American intellectuals also believed that popular fiction could "pollute" the imaginations and actions of America's vulnerable young women.[4]

Although it is impossible to determine the exact circulation figures or readership of this early American fiction, surviving figures and surveys of libraries, as well as diaries from the early Republic, seem to support the intellectuals' repeated assertions that these pieces were largely written by and for women. One recent survey of more than a thousand extant copies of the one hundred American novels published between 1789 and 1820 found women's signatures of ownership to outnumber men's by about two to one. This was not the pattern for any other kind of book in the period. Expanding female literacy, particularly in New England and the Middle Atlantic states, led novelists and their publishers to target women in their advertisements. Female readership also extended far beyond ownership at this time: women were more likely than men to share fiction at "social libraries" (266 of which were founded between 1791 and 1800) or to rent books cheaply at commercial "circulating libraries" (which more than tripled over the same decade). Moreover, as middle-class women read novels aloud to each other in spinning circles and quilting bees, listening servants and children also could be seduced by fiction's charms.[5]

Or so reasoned those intellectuals who feared that popular culture's moral contagion would spread from literate women to other simple, innocent, and vulnerable social groups. Filled with visions of luxury, ease, and sentimental attachments, America's yeoman farmers and servants might follow the women into moral decline. Royall Tyler, a lawyer in Boston and Vermont, detected signs of this decline in New England by 1797. Observing with surprise "the extreme avidity, with which books of mere amusement were purchased and perused by all ranks of his countrymen," he was particularly struck by the extent to which simple Yankee yeoman farmers had forsaken "the sober ser-

    [4]"Novel Reading, a Cause of Female Depravity" (1797), reprinted in the *New England Quarterly* (Boston, 1802), 3:172–74; "The Character and Effects of Modern Novels," *Columbian Magazine* (Philadelphia, 1792), 6:25, and *Weekly Magazine* (Philadelphia, 1798), 1:185.
    [5]See Cathy N. Davidson, *Revolution and the Word: The Rise of the Novel in America* (New York, 1986), especially chaps. 1, 2, and 4.

mons and Practical Pieties of their fathers." This turn to popular culture in the countryside boded ill for the region's traditional discipline and industry:

> The worthy farmer no longer fatigued himself with Bunyan's Pilgrim up the "hill of difficulty" or through the "slough of despond"; but quaffed wine with Brydone in the hermitage of Vesuvius, or sported with Bruce on the fairy land of Abysinia: while Dolly, the dairy maid, and Jonathan, the hired man, threw aside the ballad of the cruel stepmother, over which they had so often wept in concert, and now amused themselves into so agreeable a terrour, with the haunted houses and hobgobblins of Mrs. Ratcliffe, that they were both afraid to sleep alone.[6]

The intellectuals' fears for traditional moral authority and social order, buried in Tyler's seriocomic lamentations as in more sober jeremiads, rose to the surface of American intellectual discourse only at moments of extreme provocation. One such moment came to William Cobbett, the ubiquitous critic of American life, in 1794 as he observed the immense popularity in America of Susanna Rowson's sentimental novels and plays.

Susanna Haswell Rowson, the daughter of a British naval officer and royal customs official, was born in England, raised in Massachusetts, and returned to England as part of a prisoner exchange in 1778. After joining a theater company in Philadelphia in 1793, she finally settled in Boston in 1796. A successful actress, playwright, novelist, and founder of a respected girls' academy in Boston, Rowson supported her ne'er-do-well husband and his family throughout her long marriage. She was nothing if not industrious — and bold. All but one of her novels were published under her own name, rather than with the discreetly anonymous note "by an American lady" that identified most of the period's fiction. Her fourth novel, *Charlotte. A Tale of Truth,* was written and published in London in 1791 and republished in Philadelphia in 1794.[7] The most popular novel in America until the mid-1850s, *Charlotte* went through more than two hundred editions and sold about forty thousand copies in its first decade. Rowson and her work

---

[6] [Royall Tyler], *The Algerine Captive; or, the life and adventures of Doctor Updike Underhill, six years a prisoner among the Algerines* (Walpole, N.H., 1797), v–vii, viii–ix. The seventeenth-century English minister John Bunyan's *Pilgrim's Progress* was among the favorite reading of early New England Puritans. Ann Radcliffe wrote popular Gothic fiction in eighteenth-century England.

[7] Document 20

were thus irresistible targets for another resident of Philadelphia, the irascible English author William Cobbett.

Cobbett's venomous attack on Rowson lies at the extreme end of American intellectual discourse, but it is representative in its linking of women, popular culture, and the collapse of social authority. Cobbett found Rowson's "spirited manner" and apparent faith in "the superiority of her sex" grating. Worse still, he found Rowson's views "well received in a country, where the authority of the wise is so unequivocally acknowledged, that the *reformers* of the *reformed church,* have been obliged (for fear of losing all their custom) to raze the odious word *obey* from their marriage service." Infuriated by Rowson's ability to feel the pulse of America, Cobbett could only utter dire forebodings of social revolution:

> I do not know how it is, but I have a strange misgiving hanging about my mind, that the whole moral as well as political world is going to experience a revolution. Who knows but our present House of Representatives, for instance, may be succeeded by members of the other sex? What information might not the democrats and grog-shop politicians expect from their communicative loquacity! I'll engage there would be no secrets then.[8]

In the minds of Cobbett and many other intellectuals, American popular culture was women's culture. Susanna Rowson's dominance gave that culture a feminist cast that, they feared, would soon undermine the obedience and deference to "the authority of the wise," which alone preserved America from social revolution and political chaos. In short, they concluded, American popular culture threatened the stability and perhaps the very existence of the Republic.

The question, then, is why an apparently didactic, conservative popular culture excited these fears. Certainly on the surface, sentimental novels and plays, masquerading as moral textbooks on the supreme value of female virginity, offered no such threat to the Republic. But a deeper reading of the content and assumptions of early American fiction reveals some basis for the intellectuals' perception of a rising danger. In two overlapping ways, popular culture did offer Americans an alternative vision of reality, with subversive implications for American society and republican ideology alike. And, to give Cobbett his due,

---

[8] [William Cobbett], *A Kick for a Bite; or, Review upon Review; with a Critical Essay, on the Works of Mrs. S. Rowson; in a letter to the Editor, or Editors, of the American Monthly Review,* 2nd ed. (Philadelphia, 1796), 20–24, 27–28.

this alternative vision centered on and challenged traditional gender assumptions.

The first of these challenges was contained in the genre's disturbing portrait of men and male values. Like Rowson's *Charlotte,* many of the popular early American sentimental novels were tales of seduction and displayed a profound distrust of men. They darkly warned women, explicitly and implicitly, to "Be circumspect, be cautious, then,/Beware of all, but most of men." In the world of this popular culture, most men were moral ciphers, dangerous as tempters unless safely aged and seemingly unable to control their lusts. Women alone possessed virtue in this novelistic universe; men, as Rowson warned her female readers, "From virtue's bright refulgent throne/With baleful hand will drag you down;/Dishonour first, then leave to mourn/Those blessings which can ne'er return." Rowson's *Charlotte* demonstrates this dark lesson dramatically, as it led a host of other popular tales in portraying the "downfall" and death of an innocent heroine at the hands of a selfish seducer and his corrupt male adviser and friend. Similarly, Hannah Foster's extremely popular novel *The Coquette* traced the complex "fall" and death of an intelligent but disarmed heroine, deceived by a charming, hypocritical seducer. Male sexuality was thus portrayed again and again as an ever-present, all-consuming vice or interest, which often destroyed the duplicitous man and always destroyed the unsuspecting woman.[9]

This portrait of male sexuality as pervasive vice led naturally to the second subversive element of American popular culture: the insistence that women be vigilant and active in the preservation of virtue. This injunction lay at the heart of American popular culture, in the worst sentimental fiction as well as the best. It was frequently offered directly (and innocently) to female readers, as novelists interrupted their narratives to accentuate the moral lesson in their own voices:

> Oh my dear girls—for to such only am I writing—listen not to the voice of love, unless sanctioned by paternal approbation; be assured, it is now past the days of romance; no woman can be run away with

[9]Susanna Haswell Rowson, *Mentoria; or the Young Lady's Friend* (Philadelphia, 1794), 1:11–12; Rowson, *Trials of the Human Heart, a Novel* (Philadelphia, 1795), 4 vols., 1:137–56, 2:3–84; Rowson, *The Fille de Chambre, a Novel* (Philadelphia, 1794), 118–45. See also Susanna Rowson, *Charlotte. A Tale of Truth* (Philadelphia, 1794), and Hannah Foster, *The Coquette* (Boston, 1797). Overall, these early American seduction tales differ from their European counterparts, as well as from later nineteenth-century American sentimental fiction, in their notable lack of happy endings and domestic bliss. Theirs was a much darker vision of reality.

contrary to her own inclination: then kneel down each morning, and request kind heaven to keep you free from temptation, or, should it please to suffer you to be tried, pray for fortitude to resist the impulses of inclination, when it runs counter to the precepts of religion and virtue.[10]

In the better fiction of the period, this lesson of women's responsibility for their own actions was reinforced dramatically through a focus on strong female characters who made choices and directed their own lives, for good or ill. This was especially necessary, since the fathers in these novels were invariably absent, weak, or avaricious and unable to protect their virtuous daughters from the deceptions and sexual vice of younger men. In a world in which young men were often selfish and vicious, and old men selfish and impotent, young women simply had to make responsible choices and actively preserve their virtue.

Only rarely did this demand for greater female activity in the private sphere spill over into an overt call for women's civic participation. One such call was issued by Susanna Rowson in a popular play of 1794—the immediate provocation for Cobbett's attack. Sharp as this attack was, Rowson's logic was sharper, as she argued (both through her manifestly political play and in her own voice in an epilogue spoken on the stages of Philadelphia and New York) that American women, free of male "licentiousness," could best preserve American virtue and liberty. Setting up an equation between public and private virtue (in which women clearly outshone men), Rowson raised her own arm for liberty against "the lordly tyrant man." Hers was both an overt political act and a direct justification for full civic participation by women.[11]

Yet Rowson's direct political challenge of 1794 was unusual. More often, American popular culture threatened the social and political order of the Republic in subtler, indirect ways by undermining the gender assumptions of traditional republicanism. Classical (and American Revolutionary) republicanism associated *virtue* with the public, male sphere and *interest* or passion with the private, female sphere. Females were necessarily apolitical "idiots" (from the Greek for "private person"), according to republican political theory, precisely because of their absorption in the passions and interests of their pri-

[10]Rowson, *Charlotte*, 2 vols., 1:38.
[11]Susanna Haswell Rowson, *Slaves in Algiers; or, a Struggle for Freedom: A Play, interspersed with songs, in three acts* (Philadelphia, 1794).

vate worlds. The American revolutionaries associated "luxury, effeminacy, and corruption" or "ignorance, effeminacy, and vice." Indeed,
their definition of *virtue* (derived from the Latin *vir,* meaning "man")
had at its root the notion of "virility," or masculinity. Republican virtue
was possible, then, only because men—unlike women—were strong,
active, independent, and capable of sacrificing private interests and
passions for the public good.[12]

Far different was the map of social reality drawn by the emerging
popular culture. Here men were frequently overwhelmed by their
passions and interests, and corruption was firmly associated with
male sexuality. Women, by contrast, were largely virtuous—strong in
the defense of virtue. When they were not, men and women alike
drowned in a sea of corruption. In terms of this vision, the justification
for male domination of civic life in the name of virtue, independence,
and strength seemed to disappear. By undermining confidence in the
virtue of American men, popular culture posed a serious threat to the
republican vision—and to the deference that upheld the traditional
social order.

Thus American popular culture's conservative text of bourgeois
respectability covered a subversive feminist subtext. With multiple levels of texts inviting different readers to resolve its internal tensions
creatively, sentimental fiction may well have liberated the imaginations
of its first American readers. Whether or not a reader embraced the
feminist subtext, the very act of interpreting and completing the narrative might empower a woman to examine her own life. Certainly, Rowson's constant use of the first and second person ("I" and "you") in her
novels invited this sort of immediate, personal response. As Cobbett
and his fellow intellectuals sensed, this was a potentially revolutionary
alternative culture.

Realizing that diatribes against novels and bans on theater were
ineffective, two American intellectuals attempted to respond more con-
structively. Both had protested against popular culture earlier, without
notable success. Now they tried a different approach. Although neither was able to defeat or even to contain American popular culture,
Hugh Henry Brackenridge and Royall Tyler did join the contest in
creative ways. Each sought to accommodate the American public's

---

[12]Linda K. Kerber, *Women of the Republic: Intellect and Ideology in Revolutionary
America* (Chapel Hill, N.C., 1980), and Ruth H. Bloch, "The Gendered Meanings of
Virtue in Revolutionary America," *Signs: Journal of Women in Culture and Society,* 13,
no. 1 (1987):37–58, discuss the gender assumptions of American Revolutionary republicanism.

taste for fiction while draining it of its subversive elements. They came closer than any of their contemporaries to making novels and plays safe for republicanism.

The didacticism of Hugh Henry Brackenridge's *Modern Chivalry: Containing the Adventures of Captain John Farrago, and Teague O'Regan, His Servant,* is unmistakable.[13] Even if a postscript to the first volume of 1792 did not declare half seriously, "I hope to see it made a school book; a kind of classic of the English language," Brackenridge's desire to instruct his countrymen leaps from every page of this extended picaresque novel. His comic American characters—loosely based on Miguel de Cervantes's deluded knight-errant, Don Quixote, and his earthy squire, Sancho Panza—were created to engage and amuse American readers. More fundamentally, Brackenridge hoped that they might illuminate the follies of democratic excess in a form less likely to offend and more likely to reach a wide audience than a learned philosophical treatise or a political pamphlet.

The political lessons that Brackenridge offered were born of bitter personal experience. Following his graduation from the College of New Jersey and brief stints as a schoolmaster and chaplain to the Continental army, Brackenridge moved in 1781 to the frontier town of Pittsburgh, Pennsylvania. There he practiced law, founded a newspaper and a boys' academy, and served in the Pennsylvania State Assembly—until he was defeated in his bid to represent his district in the state's ratifying convention for the U.S. Constitution in 1787. As was often the case in western Pennsylvania, class and ethnic tensions were as important as political ideology, as his Irish opponent (an ex-weaver and Antifederalist) tarred the Scottish-born Brackenridge as an Eastern elitist and cosmopolitan Federalist. Labeling the victor a "Teague O'Regan" (the common contemporary epithet for an ignorant Irish immigrant), Brackenridge publicly ridiculed his former constituents' judgment, thus sealing his unpopularity and political exile. Throughout the 1790s, he concentrated on rebuilding his "lost" law practice and criticizing the frontier democrats through satire.

Originally, Brackenridge cast his satire in the form of a long poem, "The Modern Chevalier." In this draft, a modern knight wandered the American countryside, observing the election and subsequent misadventures of Traddle the weaver. When his frontier neighbors were not impressed by his limp verse, Brackenridge recast his political lessons in prose. His satire remained sharp: the four volumes of *Modern*

---

[13]Document 21

Chivalry published in the 1790s trace the exaltation of an illiterate Irish servant in politics, the ministry, high society, the theater, and scientific circles. Meanwhile, his master, an educated if querulous country gentleman, attempts to restrain the man's soaring ambitions and retain the services of his "bog-trotter," while voicing his (and the author's) wonder at the credulity of the American people. Undoubtedly reflecting Brackenridge's most pressing concern in 1792, a biting satire on American elections opens each of the first two volumes of the novel.

Brackenridge's decision to embed his political lessons in a narrative and especially in a novel is significant. *Modern Chivalry's* fictions and characterizations are thin, but their mere presence is remarkable in the work of an author who profoundly distrusted the sentimental romances of his day. More than a capitulation to popular taste, Brackenridge's turn to fiction seems an attempt to co-opt and redirect a dangerous genre to useful political purposes. If the novel could not be banned in America, *Modern Chivalry* suggests, perhaps its seductive narrative power could be harnessed to educate Americans about the responsibilities of citizenship in a democratic republic.

As Brackenridge attempted to transform popular culture to instruct his neighbors on the frontier, Royall Tyler aimed at a more cosmopolitan audience. While in New York City on Massachusetts government business in March 1787, this Harvard graduate, lawyer, and army officer visited the John Street Theater and met the company's star comedian, Thomas Wignell. Tyler enjoyed Wignell's performance in Richard Sheridan's *School for Scandal* but disliked the English play's emphasis on conformity to the behavior of high society and ridicule of the lower class. He decided that, with a few days' work, the form of the eighteenth-century English comedy of manners could be adapted for more edifying purposes. *The Contrast,* produced at the John Street Theater on April 16, 1787, became the first original play by an American citizen to be performed by a professional American theater company.[14] As it played on all the major stages in New York, Philadelphia, Baltimore, Boston, and Charleston and was printed by subscription in Philadelphia in 1790, Tyler's play reached the Republic's most sophisticated (and, some critics feared, most dissolute) urban audiences.

From its patriotic prologue to its moralistic conclusion, *The Contrast* offers a striking contrast to its English model. It, too, is a comedy, but its serious republican lessons are never far from the surface.

[14]Document 22

The central action of the play contrasts the duplicitous behavior of Billy Dimple with the virtuous conduct of Colonel Henry Manly. Dimple is a New York man of fashion who lost his moral moorings on a European tour and now slavishly follows the advice of the English mannerist Lord Chesterfield. Manly, a New England veteran of the Revolutionary War, trusts his own experience as well as his model, "our illustrious WASHINGTON." In New York City to petition the federal government to relieve his fellow veterans, Manly displays all the virile independence, honor, and rectitude that his name suggests. Like the Puritan ancestors he venerates, he never misses an opportunity to deliver a sermon against luxury and moral corruption, even while promenading on New York's fashionable mall. As his "sprightly" sister moans, "His conversation made me as melancholy as if I had been at church."[15]

To offset and accentuate this high seriousness, the play introduces a low comic echo of the central contrast. The subplot sets the figure of Jessamy, Billy Dimple's "genteel" English servant, against Jonathan, the "true born Yankee American son of liberty" who waits on Colonel Manly. Played to the hilt by Thomas Wignell, Brother Jonathan was destined to become a stock character in American comedy. This innocent Yankee bumpkin resists all the temptations of fashionable New York City simply by misunderstanding them; he doesn't even realize that he has been to the theater! After unsuccessfully attempting to woo a New York girl by singing "Yankee Doodle," Jonathan declares his preference for home: "Gor! she's gone off in a swinging passion, before I had time to think of consequences. If this is the way with your city ladies, give me the twenty acres of rock, the Bible, the cow, and Tabitha, and a little peaceable bundling." Clearly hoping that its urban American audiences would draw the same conclusion either from Jonathan's comic antics or Manly's weighty sermons, The Contrast offered republican lessons wrapped in the form of popular culture.[16]

The language, structure, and prologue of The Contrast all entreat its audience to choose "homespun" American values over "ready-made" English manners. The play defines American identity in opposition to English culture. Indeed, Colonel Manly's insistence that neither foreign travel nor foreign books should shape America's new republican man (belied to some degree by his long soliloquy on ancient Greece) seems to clash with Captain Farrago's equally strong demand in Mod-

---

[15]Royall Tyler, The Contrast (Philadelphia, 1790), 20, 19, 18.
[16]Ibid., 26, 42, 47.

ern Chivalry that America's republican citizens defer to classically educated legislators. The play and the novel both present lessons in republicanism, though from somewhat different political positions and for different audiences.[17]

The Contrast and *Modern Chivalry* also share a fundamental gender orientation. As Brackenridge delineates a republican citizenship in which women could not participate and Tyler constructs a republican masculinity that women could only admire, they are clearly speaking to and privileging men. Masculine virtue, they argue, will determine American identity and the future of the American Republic. Rowson's "dear girls" are entirely displaced as the preferred audience and primary subjects of the fiction, as of the Republic. In dismissing the dichotomy of female virtue and male vice that lay at the heart of *Charlotte* and the other seduction novels of the period, Tyler and Brackenridge reversed the gendered message of American popular culture even as they appropriated its forms.

[17]A decade later, Tyler moved closer to Brackenridge's stance. His male-dominated picaresque novel of 1797, *The Algerine Captive,* deplores the dangerous and growing gap between America's educated elite and its ignorant masses—due in part to the general appetite for sentimental fiction.

6

Encountering the Other

As they attempted to define and shape American identity through a national culture, the intellectuals of the early Republic discovered their need for "the Other." They found it impossible to think and write about American identity without also thinking and writing about its negative image, or everything that it was not. Civilization could only be described in relation to savagery, virtue to vice, freedom to slavery, and so forth. For the sake of cognitive coherence, American intellectuals tended to focus and project most of those negative qualities on a single social group. American identity, like other social identities, was constructed in contrapuntal fashion, in opposition to a culturally different Other.

As poststructuralist theorists in a variety of disciplines have found in recent years, the Other may be marked by differences in race, gender, class, religion, ethnicity, or a combination of cultural categories. These categories of identity appear to be real to their makers. They may seem to be divinely ordained or natural, determined by biological or psychological heredity. Over time, they acquire the status of knowledge, even of self-evident fact. Individuals within the dominant social group believe that they know what race and gender mean and what they have always meant. This fixity is precisely their strength: once established, the cultural categories lend stability and justification to profoundly unstable and contestable identities and relationships of power.

Yet this sense of permanence and immutability is illusory. The categories of identity are not given, but constituted through learned and popular discourses of encounter. A central role in the process of "othering" belongs to narrative, or the stories that people tell about themselves and others. Even material issues between the self and the Other are shaped by narrative. For example, the battle might be over land, one scholar notes, "but when it came to who owned the land, who had the right to settle and work on it, who kept it going, who won it back, and who now plans its future—these issues were reflected, contested, and even for a time decided in narrative."[1] Inscribed and reinscribed in elite and popular culture, these stories attain the indisputability of myth.

Encountering the Other thus becomes a crucial element in the social construction of identity. But it is not an easy process. For the Other must seem not only terrible or inferior but also alluring. The Other attracts as well as repels; it is dangerous precisely because of this duality. The danger in facing the Other is not just external. Rather, the Other threatens the integrity of the self by offering alternative, unrealized, and suppressed possibilities. The element of domination in the discourse of the Other may derive in part from the felt need to repress these alternative cultural visions and to maintain belief in a stable, fixed, essential identity. For most people, then, the cultural challenge in an encounter with the Other lies in the ability to misunderstand, to face down the danger of fragmentation or division, and to reinforce the coherence of the constructed self. For a few brave souls, the deeper challenge in such an encounter might be to confront the instability of all identities—that "despite our desperate, eternal attempt to separate, contain, and mend, categories always leak"—and to consider repressed alternatives to their own culture.[2]

Without either an American identity or an Other in place, the intellectuals of the early Republic were free to create a variety of oppositions. Hugh Henry Brackenridge had his Teague O'Regan, and William Cobbett his feminist women. Of the many possible categories of otherness, however, one appears with enough frequency and force to deserve special consideration. Despite the persuasive power of gender, ethnic, and class differences in this period, race was already becoming the most significant cultural category. For many American thinkers in the early Republic, the Indian was the Other. Their confidence in the future of American "civilization" depended in large part on the construction of a "savage" Indian Other.[3]

---

[1] Edward W. Said, *Culture and Imperialism* (New York, 1993), xii–xiii.

[2] Trinh T. Minh-ha, *Woman, Native, Other: Writing Postcoloniality and Feminism* (Bloomington, Ind., 1989), 94.

[3] Because the Other must seem attractive and free, as well as dangerous and wild, the African was rarely portrayed as the Other in this period. Degraded in slavery and restrained even in freedom, the African in Revolutionary America did not represent an alluring alternative identity to the intellectual elite. This is not to say, of course, that virulent racism was not directed against Africans in American intellectual discourse. For a striking illustration of this distinction, see Thomas Jefferson, *Notes on the State of Virginia* (Philadelphia, 1788).

To do its cultural work for its creators, the image of the Indian had to be flattened, reduced, simplified, and frozen. The variety and complexity of Indian cultures was effaced, creating a simple picture of savagery and primitivism. Even when American writers defended the Indians' potential for improvement or detected the noble savage among the many ignoble savages in their path, they still believed in the essential inferiority of primitive to civilized life. With nature as their only guide, Indians could be brave and noble at one moment and fiercely savage the next, these writers suggested. Together, they constructed a savage Other—devoid of cultural complexity, integrity, and history—to define a very different American identity.

Immediately behind this image lay the most celebrated European contribution to the intellectual discourse on Indians, the *Histoire Naturelle* (1749–89) of the French naturalist Comte Georges Louis Leclerc de Buffon. Buffon's text worked in several ways to dehistoricize Indians. First, it argued that the American environment was so primitive and noxious that it could sustain only weak, underdeveloped, listless "savages." Then it offered a biological explanation of the Indians' imagined lack of sociability and morality: the small sex organs of Indian males, Buffon argued, led to universally frozen hearts, cold societies, and cruel empires. These deterministic arguments established the naturalizing, timeless, reductionist shape of the European narrative.[4]

The European philosophes who echoed and popularized Buffon's views fully released the dehumanizing tendencies of this naturalized representation. Abbé Guillaume Thomas François Raynal and Abbé Corneille de Pauw, who wrote the authoritative article on "l'Amérique" for the *Supplément* to Denis Diderot's *Encyclopédie,* relied on and extended Buffon's interpretation. De Pauw expressed no wonder that, lacking intelligence, passion, civilization, and perhaps even souls, "at first Americans were not thought to be men, but rather Orang-Outangs [orangutans], or large monkeys, that could be destroyed without remorse and without reproach." Raynal carried the analysis down to the eighteenth century, noting that "America has not yet pro-

---

[4]Comte Georges Louis Leclerc de Buffon, *Histoire Naturelle, générale et particulière* (Paris, 1761), 9:104–11. The most important volumes of Buffon's *Histoire Naturelle* were published serially between 1749 and 1767. Nine volumes on birds, five on minerals, and seven supplementary volumes came out between 1770 and 1789, the year following Buffon's death. The complete set of forty-four volumes was issued in Paris in 1804.

duced one good poet, one able mathematician, one man of genius in a single art or a single science."[5]

These sweeping European portraits of savage Indians provoked impassioned and complicated reactions from many American thinkers. The frequent conflation of the categories "American" and "savage" in these European texts—and their clear implication that European immigrants might also degenerate in the "primitive" American environment—spurred some aspiring American naturalists to a dual response. They sought at once to refute the charge of the total degeneracy of Indians (and the American environment) and to distance themselves from these "savages." That both goals could be met at once is apparent in the most famous response, Thomas Jefferson's *Notes on the State of Virginia.*[6]

Written in response to a questionnaire about the American states circulated by the French legation in 1780, Jefferson's *Notes* began as a statistical survey and expanded into the most comprehensive statement of his philosophical, social, political, and scientific beliefs. It was also his only published book. Its broad critique of the tyranny of the church, the rich and powerful, and other traditional impediments to individual freedom and happiness expands on and realizes the philosophical implications of Jefferson's famous proclamations in the Declaration of Independence.

Yet the publication of his *Notes* in London in 1787 and in Philadelphia the following year was as much a cause of anxiety as of celebration for Jefferson, who feared that the book might undermine his efforts to secure the gradual emancipation of slaves in Virginia. Indeed, Jefferson's complex views on race and slavery are nowhere more fully revealed. Controversial in his own time and ever since, Jefferson's dual belief in the injustice of slavery and the ineradicable inferiority of blacks established the limits of his faith in natural human equality.

Still more tortured, and much longer and more detailed, is Jefferson's treatment of Indians in *Notes.* Although the book resoundingly rejects the prevailing European Enlightenment theory of the degeneracy of the American environment and its native (and transplanted)

[5]Abbé Corneille de Pauw, *Recherches philosophiques sur les Américains* (Berlin, 1768), 1:35–36; Abbé Guillaume Thomas François Raynal, *Histoire Philosophique et Politique des Établisement et du Commerce des Européens dans les deux Indes* (Amsterdam, 1770), 92.

[6]Document 23

inhabitants, it also adopts the Enlightenment stance of distancing Indians as an object of study and classification. Significantly, the narrative structure places Indians within natural rather than human history. They appear under the heading "Animals," intermingled with descriptions of mammoths and birds, fish and insects. This choice, shaped by the European Enlightenment theories that Jefferson wished to refute, also reflects his regret that "the races of black and of red men . . . have never yet been viewed by us as subjects of natural history."[7]

Notes repeatedly affirms this placement of Indians among animals rather than men. To contest the European charge "that our country, from the combined effects of soil and climate, degenerated animal nature, in the general, and particularly the moral faculties of man," Jefferson invokes the Mingo chief John Logan's *natural* eloquence, as well as the *cultivated* genius of George Washington, Benjamin Franklin, and David Rittenhouse (American astronomer and mathematician). The text also naturalizes Indians by insisting that they have "never submitted themselves to any laws, any coercive power, any shadow of government. Their only controuls are their manners, and that moral sense of right and wrong, which, like the sense of tasting and feeling, in every man makes a part of his nature." Jefferson's determination to "open and examine" a sacred Indian burial ground, still visited by Indians, in the interests of science also vividly illustrates his tendency to objectify and distance these "barbarous people."[8]

Overall, Jefferson's defense of the Indians' physical bravery and eloquence fails to disturb his conclusion that "there are varieties in the race of man, distinguished by their powers both of body and mind," or his desire to keep the races of man "as distinct as nature has formed them." No doubt is ever raised about which is the superior and normative race. In language and structure, the text maintains a fundamental difference between "civilized" Euro-Americans and the "savage" strangers they encounter in the wilderness. Due to the fame of its author, as well as the book's elegance and obvious conviction, *Notes on the State of Virginia* played a large role in constructing an image of the savage Other in American culture.[9]

If Jefferson's *Notes* endows its "savage" Indians with physical courage and eloquence, Timothy Dwight's *Greenfield Hill* of 1794 is less generous.[10] In the fourth part of this pastoral epic poem, the con-

[7]Jefferson, *Notes on the State of Virginia* (Philadelphia, 1788), 153.
[8]Ibid., 67–68, 99–100, 103, 64.
[9]Ibid., 68, 153–154, 100.
[10]Document 24

servative Connecticut minister's verse becomes bloody and enraged as it narrates a markedly one-sided version of the Puritan-Pequot war of 1637. Ignoring the land-grasping, vengeful, and merciless aspects of the Puritans' campaign against Pequot men, women, and children, the poem portrays the Puritans as innocent victims of a Pequot conspiracy to destroy them. As true Christians, Dwight's Puritans offer their attackers "sweet smiles," faith, and friendship, only to encounter the vindictive rage of "remorseless Indians" who "scalp'd the hoary head, and burn'd the babe with fire."[11]

Dwight's animal-like Pequots, ferocious savages all, respond with indiscriminate "slaughter" that reveals their basic inhumanity: "Fierce, dark, and jealous, is the exotic soul,/That, cell'd in secret, rules the savage breast./There treacherous thoughts of gloomy vengeance roll,/And deadly deeds of malice unconfess'd;/The viper's poison rankling in it's [its] nest." The utter "destruction" of these dark, unthinking, and unrepentant "fiends of blood," the poem concludes, was both deserved and necessary for the advance of civilization. Inflammatory in language and tone, *Greenfield Hill* not only rewrote history in favor of the victors but also inscribed an image of the fierce, savage Other into early American culture.[12]

This representation took visual form in John Vanderlyn's celebrated painting of 1804, *The Death of Jane McCrea*.[13] A native of New York, Vanderlyn worked for a while as the painter Gilbert Stuart's assistant in Philadelphia. In 1796 he traveled to Paris, where he studied at the École des Beaux-Arts and completed a series of history paintings. The first of these, the *Death of Jane McCrea*, is based on an actual event that received widespread attention during and after the American Revolution. Jane McCrea, an American woman traveling through enemy lines to meet the English army officer she intended to marry, was murdered by two Mohawk warriors in July 1777. Her death was remembered and rehearsed for many decades in popular ballads and poems, as well as in the American press.

The painting followed earlier portrayals of McCrea as the personification of innocence. If the viewer's eye is drawn first to McCrea, however, her own upward stare directs all eyes toward the cruel face (and

---

[11]Timothy Dwight, *Greenfield Hill: A Poem in Seven Parts* (New York, 1794), part 4, 93–105, lines 329, 357, 123–26.

[12]Ibid., 325, 316–20, 237. For a striking contrast to Dwight's treatment of the Puritan-Pequot war, see Francis Jennings, *The Invasion of America: Indians, Colonialism, and the Cant of Conquest* (Chapel Hill, N.C., 1975).

[13]Document 25

ax) of her murderer. The painting's focus on the demonic ferocity of the Mohawks, who show no mercy for the lone, defenseless woman in their power, dramatically constructs a visual image of savagery. By freezing the scene at the moment before the murder, *The Death of Jane McCrea* invites viewers to participate over and over again in the terror of facing the savage Other.

Yet this representation of the Indian, though pervasive, was not universal among American intellectuals. There was at least one important exception to the rule. William Bartram, the Philadelphia naturalist who published *Travels through North & South Carolina, Georgia, East & West Florida, the Cherokee Country, the Extensive Territories of the Muscogulges or Creek Confederacy, and the Country of the Chactaws* in 1791, struggled with and against the conventional narrative of encounter.[14]

William Bartram was well aware of—we might say fully educated in—the representation of the savage Indian Other. On a personal level, his father, the Quaker naturalist John Bartram, felt a lifelong revulsion for Indians after his own father's death at their hands. More broadly, William Bartram had ample exposure to the cultural constructions of his contemporaries and was well connected to the leading intellectuals of his day. He maintained a wide correspondence, entertained a steady stream of prominent visitors to his father's botanical garden, and was elected to an early membership in the American Philosophical Society of Philadelphia. He was invited by President Thomas Jefferson to serve as naturalist on a government-sponsored expedition up the Red River in 1803. (Bartram declined because of advancing age and ill health.) Clearly, Bartram was an insider, an integral part of the nation's most vital intellectual community in the early Republic.[15]

His position as an insider notwithstanding, Bartram moved subtly but decisively beyond the dominant representation of the Indian as savage Other. His *Travels* began to depict Indians in a new way, reflecting a deeper appreciation for the complexity, diversity, and integrity of Indian societies. In his prolonged encounter with Indians, Bartram found not only similarities he could love but also differences he could respect. By looking long and hard, he saw past the savage Other to a complicated alternative culture.

[14]Document 26

[15]Among William Bartram's visitors after his father's death in 1777 were George Washington, George Mason, James Madison, Alexander Hamilton, the poet Hugh Williamson, the artist William Dunlap, the novelist Charles Brockden Brown, and the ornithologist Alexander Wilson. See *The Travels of William Bartram,* ed. Francis Harper (New Haven, Conn., 1958), xviii, xxviii–xxxi.

Bartram's narrative operates on several levels at once as it negotiates between distinct traditions. Carefully and even poetically written, his *Travels* is consciously literary. His intricate descriptions of flora and fauna impressed and influenced the English Romantic poets William Wordsworth and Samuel Taylor Coleridge. The book also records and reinvents Bartram's actual four-year journey through the Carolinas and Floridas, Georgia, and the lands of the Cherokees, Muskogees, and Choctaws twenty years earlier. This expedition, sponsored by Dr. John Fothergill, a British naturalist and patron of John Bartram, retraced the latter's travels and realized some of his dreams of naming and collecting botanical samples in new areas. Designed "to search the Floridas, and the western parts of Carolina and Georgia, for the discovery of rare and useful productions of nature, chiefly in the vegetable kingdom," William Bartram's quest contributed to the Enlightenment project of naming, classifying, categorizing, knowing, and using the wonders of nature.[16]

Like Jefferson's *Notes on the State of Virginia*, Bartram's *Travels* initially places Indians among the plants and animals in its catalog of natural history. But Bartram's uneasiness with this placement and its naturalizing and dehumanizing effects erupts throughout the book. Shaped by the dominant representation of the savage Other, Bartram's narrative struggles to make the prevailing image fit his own experience of encounter. Although the text never completely escapes the available language of savagery and civilization, it constantly questions and undermines its basic assumptions. Bartram's Indians think, speak, argue, show mercy, and grapple with the consequences of the contact of cultures. They often know more than the Anglo-American settlers and traders they encounter. They have rich cultures, histories, and moral codes. In short, they are self-conscious, sophisticated human beings.

A highly dramatic episode early in *Travels* forms a striking contrast to Vanderlyn's *Death of Jane McCrea* and Dwight's *Greenfield Hill*. Traveling alone and unarmed on horseback one evening, Bartram meets an armed and apparently hostile mounted Seminole. The unexpected encounter initially evokes for him all the fears taught him by his culture and especially his father. Unable to hide, Bartram prepares to die, commending his soul to God. But suddenly his expectation of ferocious savagery abates sufficiently for him to attempt to communicate with the Other:

[16]William Bartram, *Travels through North & South Carolina, Georgia, East & West Florida, The Cherokee Country, the Extensive Territories of the Muscogulges or Creek Confederacy, and the Country of the Chactaws* (Philadelphia, 1791), 1.

The intrepid Siminole stopped suddenly, three or four yards before me, and silently viewed me, his countenance angry and fierce, shifting his rifle from shoulder to shoulder, and looking about instantly on all sides. I advanced towards him, and with an air of confidence offered him my hand, hailing him, brother; at this he hastily jerked back his arm, with a look of malice, rage and disdain, seeming every way disconcerted; when again looking at me more attentively, he instantly spurred up to me, and, with dignity in his look and action, gave me his hand.[17]

Bartram is not content merely to present this dramatic narrative. Thinking aloud, searching for its meaning, he offers an interpretation of this wonderful change:

Possibly the silent language of his soul, during the moment of suspense (for I believe his design was to kill me when he first came up) was after this manner: "White man, thou art my enemy, and thou and thy brethren may have killed mine; yet it may not be so, and even were that the case, thou art now alone, and in my power. Live; the Great Spirit forbids me to touch thy life; go to thy brethren, tell them thou sawest an Indian in the forests, who knew how to be humane and compassionate." In fine, we shook hands, and parted in a friendly manner, in the midst of a dreary wilderness; and he informed me of the course and distance to the trading-house, where I found he had been extremely ill treated the day before.[18]

Bartram's decision to construct an imaginary dialogue between himself and the Other is highly significant in transforming the scene. Although it cannot give actual voice to the Seminole, it does create the sense of communication between two equal, thinking human beings. It also balances and provides a context for Bartram's next, actual dialogue with the white traders nearby. In equally implausible language, the "chief" trader reports,

with a countenance that at once bespoke surprise and pleasure, "My friend, consider yourself a fortunate man: that fellow," said he, "is one of the greatest villains on earth, a noted murderer, and outlawed by his countrymen. Last evening he was here, we took his gun from him, broke it in pieces, and gave him a severe drubbing: he, however, made his escape, carrying off a new rifle gun, with which, he said, going off, he would kill the first white man he met."[19]

[17]Ibid., 21.
[18]Ibid.
[19]Ibid., 22.

Bartram's interpretation of this incident decisively reverses the prevalent representation of civilized whites and savage Indians. He concludes twice that the Seminole had been ill-treated by the traders and yet managed to rise above a simple, vengeful response. Making fine distinctions between the traders he knows and the stranger he meets, the Seminole had thought his way through a complicated moment in the contact of cultures. He had shown judgment, mercy, and humanity. Overall, the text concludes, the "savage" had displayed more intelligence, restraint, and dignity than the "civilized."[20]

At this point in the narrative, Bartram attributes this surprising inversion to an innate moral instinct or divine light operating within the Seminole. "Can it be denied, but that the moral principle, which directs the savages to virtuous and praiseworthy actions, is natural or innate?" he asks. Indeed, he reasons, unschooled as they are in "philosophy, where the virtuous sentiments and actions of the most illustrious characters are recorded, and carefully laid before the youth of civilized nations," the Indians must rely on instinct. "This moral principle must be innate, or they must be under the immediate influence and guidance of a more divine and powerful preceptor, who, on these occasions, instantly inspires them, and as with a ray of divine light, points out to them at once the dignity, propriety, and beauty of virtue."[21]

Except for the allusion to the Quaker concept of a divine Inner Light, this explanation might have come from Jefferson's *Notes on the State of Virginia*. But Bartram's apparent celebration of primitive virtue and noble savagery, presented in the standard language of the dominant discourse ("savages" opposed to "civilized nations"), has already been undermined by the contours of the episode just narrated. That detailed account of a complex contact of cultures, set not in a distant past but in a historical present marked by dialogue and conflict, belies a simple primitivist interpretation. That Bartram presents both possibilities to his readers indicates his difficulty in reconciling his society's representations of the savage Other with his own, more complicated experience of encounter.

This early scene places an Indian in the wilderness but allows him dignity and humanity. The fourth and final part of *Travels* takes the still more radical step of separating Indians from the wilderness. In a significant structural departure from the norms of natural history,

[20] Ibid., 21–22.
[21] Ibid., 22–23.

Bartram's "Account of the Persons, Manners, Customs and Government, of the Muscogulges, or Creeks, Cherokees, Chactaws, &c. Aborigines of the Continent of North America," allows Indians to stand alone, as human beings and historical subjects. Distinguishing between tribal cultures and according each its own history, the narrative subverts the dominant construction of Indians as objects in nature.

At times, Bartram seems conscious of this subversion, as when he insists on the virtue of the Indians he has met:

> If we consider them with respect to their private character or in a moral view, they must, I think, claim our approbation, if we divest ourselves of prejudice and think freely. As moral men they certainly stand in no need of European civilization.
>
> They are just, honest, liberal and hospitable to strangers; considerate, loving and affectionate to their wives and relations; fond of their children; industrious, frugal, temperate and persevering; charitable and forbearing. I have been weeks and months amongst them and in their towns, and never observed the least sign of contention or wrangling: never saw an instance of an Indian beating his wife, or even reproving her in anger. In this case they stand as examples of reproof to the most civilized nations, as not being defective in justice, gratitude and a good understanding. . . .[22]

This frank statement of respect and even preference for Indian morality, embedded in a careful description of the variety of tribal cultures, undermines the conventional representation of the savage Indian. So, too, does the drawing of "Mico Chlucco the Long Warrior or King of the Siminoles," looking dignified and solemn, that serves as the frontispiece of *Travels*. Attesting to the special quality of Bartram's prolonged encounter with the Other, they prepare the way for his provocative final challenge: "Do we want wisdom and virtue? let our youth then repair to the venerable councils of the Muscogulges."[23]

Thus, even as the image of the savage Other was being constructed in the early Republic, it was also being quietly deconstructed by one intrepid intellectual adventurer. Bartram's subtle subversions did not overturn his fellow intellectuals' discourse of domination. But he did present an alternative narrative of encounter and openly invited his own and subsequent generations to follow him on his journey to a fuller appreciation of cultural complexity and diversity in America.

[22] Ibid., 489–90.
[23] Ibid., 493.

7

Conclusion

Even their most fervent admirers would have to admit that the founders of the American state did not succeed in their quest to create a strong, unified, national culture between 1775 and 1800. Their efforts to invent an American language came to naught; their epic poetry was unread and largely unreadable. Despite Washington's bequest and Jefferson's efforts, no national university was ever established. Nor were the intellectuals' many plans for a national system of public education realized in this period. The histories that hoped to narrate nationhood and unify a divided society were read as politically or regionally partisan. Waging a losing battle against popular culture and encountering the Other undermined some intellectuals' confidence and clarity of vision.

Meanwhile, in Philadelphia, the center of the American Enlightenment and the seat of national government, a series of devastating yellow fever epidemics divided and dispirited the vital intellectual community in the 1790s. As their many admirers and correspondents around the country watched in horror, Benjamin Rush and his fellow physicians bickered over causation and treatment while political leaders and citizens alike fled the capital. By 1800, many American intellectuals were scarred and weary.[1]

Succeeding generations faced challenges beyond the ken of the Revolutionary intellectuals. In the North, a new web of market relations allowed fewer and fewer Americans the economic and social independence of the republican ideal. Partisanship was inscribed permanently into American political culture, and the federal union barely survived increasing regional tensions. A new wave of evangelical religious revivals swept the country, offering individual redemption to those who attended to the state of their private souls. A national

---

[1] See Eve Kornfeld, "Crisis in the Capital: The Cultural Significance of Philadelphia's Great Yellow Fever Epidemic," *Pennsylvania History,* 51, no. 3 (1984): 189–205.

culture based on civic virtue seemed ever more remote and impossible to create.

Equally destructive of the intellectuals' dream of an independent American culture, the end of decades of war in 1815 led growing numbers of elite Americans to visit Europe again. The worst fears of the Revolutionary generation were realized, as these young Americans discovered the powerful new ideals of individual development and apolitical culture that had been circulating in Europe for more than half a century. In the decades that followed, eager American travelers returned home to introduce Swiss and German theories of liberal education and self-development, German historicism and scholarly detachment, and the English Romantics' celebration of individual imagination and private visions of reality—all inimical to a republican national culture.

The insularity of the Revolutionary intellectuals' cultural vision thus seemed particularly limited to later generations of American intellectuals. As the fear of American political collapse and foreign contagion receded after 1815, American intellectual leaders came to see more excitement than danger in European culture. To them, the American revolutionaries' view of the world appeared too small, cautious, and provincial. Gone were the days when American intellectuals could dismiss the German philosopher Immanuel Kant's entire "system of metaphysics and moral philosophy" as "obscure and scarcely intelligible" and probably subversive of "all religion and morals" because it could not be translated into plain English. Nearly four decades later, the American Transcendentalist Ralph Waldo Emerson reversed the revolutionaries' judgment, proclaiming the hopeless immaturity and obscurity of American culture.[2]

Yet this central project of the Revolutionary intellectuals did not disappear without a trace. Those elements that best met the need of an increasingly democratic, capitalist, evangelical society for basic instruction and uncomplicated heroes survived through the generations and the centuries. American children of several generations learned to read, spell, and "lisp the praise of liberty" and American heroes from Noah Webster's *American Spelling Book*.[3] Their parents

[2]Samuel Miller, *A Brief Retrospect of the Eighteenth Century* (New York, 1803), 2 vols., 2:26–27.

[3]Noah Webster, *On the Education of Youth in America* (Boston, 1970), reprinted in *Essays on Education in the Early Republic,* ed. Frederick Rudolph (Cambridge, Mass., 1965), 65.

relied on Webster's dictionary. Together, they stared at copies of Gilbert Stuart's portraits of George Washington, mounted on the walls of public schools founded at mid-century by another set of educational reformers. Images of the "American Colonist" and the American patriots created by David Ramsay and Mercy Otis Warren shaped their view of the past. Perhaps unconsciously, they also absorbed the representations of a gendered, ethnic, and racial Other constructed by the intellectuals of the early Republic. And few Americans of any generation escaped Mason Locke Weems's mythological narrative of George Washington's youth.

The campaign to create an American culture between 1775 and 1800 also left its mark in subtler, less tangible ways. Quite unintentionally, the intellectuals of the Revolutionary generation initiated a New Englandization of American culture that lasted well into the twentieth century. Although regional dialects and values did not disappear, American high culture was identified with New England ever after Noah Webster, Mercy Otis Warren, Royall Tyler, and other Revolutionary intellectuals insisted that New England form the standard for American speech, history, and morality. Even David Ramsay of South Carolina contributed to this phenomenon, for his mythical American Colonist bore a remarkable resemblance to a New England yeoman farmer.

Once established, this attitude became both pervasive and invisible. Southern and western culture might be acknowledged to exist, but they were inevitably defined as regional cultures; New England culture was considered American culture. During the next American cultural renaissance of the mid-nineteenth century, the New England Transcendentalists unconsciously adopted this stance. Ralph Waldo Emerson, Margaret Fuller, and their intellectual circle might have railed against the provincialism of the Revolutionary intellectuals' fear of European culture, but their own unexamined desire to spread New England values and learning around the country marked them as equally provincial—and excellent students of the Revolutionary generation.

In the end, their commitment and persistence in the face of insurmountable difficulties seem more striking than the Revolutionary intellectuals' failure to create a unified national culture. Although they could not heal long-standing political divisions or quiet political contentiousness, they did build a vital intellectual community that stretched from Maine to South Carolina and survived vicious partisan strife. Although they could not unify a heterogeneous society or overcome more than a century of cultural diversity, they did open a conversation about American identity that has never ended. Even those entirely excluded from the original conversation could not resist joining it. The Revolutionary intellectuals' vision of American cultural unity, like the myth they created of George Washington's cherry tree, proved remarkably enduring, even in democratic, materialistic, pluralistic, modern America.

PART TWO

The Documents

A Note about the Text:
Written English in the last quarter of the eighteenth century differs somewhat from present-day written English. Since orthography was a central concern of Benjamin Franklin, Noah Webster, and other American authors of the period, original spellings and punctuation are maintained throughout the documents in this volume. In the interest of clarity and accessibility to the reader, however, the traditional long "s" (f) has been silently modernized.

In the following documents, numbered footnotes are those provided by the editor of this volume; asterisked footnotes are part of the original document.

Inventing an American Language and Literature

1

PHILIP FRENEAU AND
HUGH HENRY BRACKENRIDGE

A Poem on the Rising Glory of America
1772

A Poem on the Rising Glory of America *was delivered at commencement at the College of New Jersey (later called Princeton) in September 1771 by two young graduates, Philip Freneau (1752–1832) and Hugh Henry Brackenridge (1748–1816). In later years, Freneau would win fame as the "Poet of the Revolution" and as a central figure in the Jeffersonian Republican party; Brackenridge would become a lawyer, legislator, justice of the supreme court of Pennsylvania, and author. Their appreciative audience included John Witherspoon, the president of the college and a future signer of the Declaration of Independence. Their classmate James Madison missed commencement due to poor health. Published in 1772 in Philadelphia, the poem was also well received in other colleges and colonies.*

Philip Freneau and Hugh Henry Brackenridge, *A Poem on the Rising Glory of America* (Philadelphia, 1772), reprinted in *The Poems of Philip Freneau,* ed. Fred L. Pattee (New York, 1963), 3 vols., 1:50–52, 58–60, 70–71, 78, 82–83.

Leander

No more of Memphis and her mighty kings,
Or Alexandria, where the Ptolomies
Taught golden commerce to unfurl her sails,
And bid fair science smile: No more of Greece
Where learning next her early visit paid,
And spread her glories to illume the world;
No more of Athens, where she flourished,
And saw her sons of mighty genius rise,
Smooth flowing Plato, Socrates and him
Who with resistless eloquence reviv'd
The spirit of Liberty, and shook the thrones
Of Macedon and Persia's haughty king.
No more of Rome, enlighten'd by her beams,
Fresh kindling there the fire of eloquence,
And poesy divine; imperial Rome!
Whose wide dominion reach'd o'er half the globe;
Whose eagle flew o'er Ganges to the East,
And in the West far to the British isles.
No more of Britain and her kings renown'd,
Edward's and Henry's thunderbolts of war;
Her chiefs victorious o'er the Gallic foe;
Illustrious senators, immortal bards,
And wise philosophers, of these no more.
A Theme more new, tho' not less noble, claims
Our ev'ry thought on this auspicious day;
The rising glory of this western world,
Where now the dawning light of science spreads
Her orient ray, and wakes the muse's song;
Where freedom holds her sacred standard high,
And commerce rolls her golden tides profuse
Of elegance and ev'ry joy of life.

Acasto

Since then, Leander, you attempt a strain
So new, so noble and so full of fame;
And since a friendly concourse centers here,
America's own sons, begin O muse!
Now thro' the veil of ancient days review

The period fam'd when first Columbus touch'd
The shore so long unknown, thro' various toils,
Famine and death, the hero made his way,
Thro' oceans bellowing with eternal storms.
But why, thus hap'ly found, should we resume
The tale of Cortez, furious chief, ordain'd
With Indian blood to dye the sands, and choak
Fam'd Amazonia's stream with dead! Or why
Once more revive the story old in fame,
Of Atabilipa,[1] by thirst of gold
Depriv'd of life: which not Peru's rich ore,
Nor Mexico's vast mines cou'd then redeem.
Better these northern realms deserve our song,
Discover'd by Britannia for her sons;
Undeluged with seas of Indian blood,
Which cruel Spain on southern regions spilt;
To gain by terrors what the gen'rous breast
Wins by fair treaty, conquers without blood.

. . .

Leander

How fallen, Oh!
How much obscur'd is human nature here!
Shut from the light of science and of truth
They wander'd blindfold down the steep of time;
Dim superstition with her ghastly train
Of daemons, spectres and foreboding signs
Still urging them to horrid rites and forms
Of human sacrifice, to sooth the pow'rs
Malignant, and the dark infernal king.
Once on this spot perhaps a wigwam stood
With all its rude inhabitants, or round
Some mighty fire an hundred savage sons
Gambol'd by day, and filled the night with cries;
In what superior to the brutal race
That fled before them thro' the howling wilds,
Were all those num'rous tawny tribes which swarm'd
From Baffin's bay to Del Fuego south,

---

[1]*Atabilipa:* Atalhualpa, (1500?–1533) last Inca king of Peru, was captured by Francisco Pizarro in 1531 and executed in 1533.

From California to the Oronoque[2]?
Far from the reach of fame they liv'd unknown
In listless slumber and inglorious ease;
To them fair science never op'd her stores,
Nor sacred truth sublim'd the soul to God;
No fix'd abode their wand'ring genius knew;
No golden harvest crown'd the fertile glebe;
No city then adorn'd the river's bank,
Nor rising turret overlook'd the stream.

Acasto

Now view the prospect chang'd; far off at sea
The mariner descry's our spacious towns,
He hails the prospect of the land and views
A new, a fair, a fertile world arise;
Onward from India's isles far east, to us
Now fair-ey'd commerce stretches her white sails,
Learning exalts her head, the graces smile
And peace establish'd after horrid war
Improves the splendor of these early times.
But come, my friends, and let us trace the steps
By which this recent happy world arose,
To this fair eminence of high renown
This height of wealth, of liberty and fame.
.   .   .

Leander

Great is the praise of commerce, and the men
Deserve our praise who spread from shore to shore
The flowing sail; great are their dangers too;
Death ever present to the fearless eye
And ev'ry billow but a gaping grave;
Yet all these mighty feats to science owe
Their rise and glory.—Hail fair science! thou,
Transplanted from the eastern climes, dost bloom
In these fair regions . . .
Hither they've wing'd their way, the last, the best

[2]*Oronoque:* Orinoco River of South America

Of countries where the arts shall rise and grow
Luxuriant, graceful; and ev'n now we boast
A Franklin skill'd in deep philosophy,
A genius piercing as th' electric fire,
Bright as the light'ning's flash, explain'd so well
By him, the rival of Britannia's sage.
This is a land of ev'ry joyous sound
Of liberty and life; sweet liberty!
Without whose aid the noblest genius fails,
And science irretrievably must die.

. . .

Leander

And here fair freedom shall forever reign.
I see a train, a glorious train appear,
Of Patriots plac'd in equal fame with those
Who nobly fell for Athens or for Rome.
The sons of Boston, resolute and brave,
The firm supporters of our injur'd rights,
Shall lose their splendours in the brighter beams
Of patriots fam'd and heroes yet unborn.

Acasto

'Tis but the morning of the world with us
And Science yet but sheds her orient rays.
I see the age, the happy age, roll on
Bright with the splendours of her mid-day beams,
I see a Homer and a Milton rise
In all the pomp and majesty of song,
Which gives immortal vigour to the deeds
Atchiev'd by Heroes in the fields of fame. . . .

This is thy praise, America, thy pow'r,
Thou best of climes, by science visited,
By freedom blest and richly stor'd with all
The luxuries of life. Hail, happy land,
The seat of empire, the abode of kings,
The final stage where time shall introduce
Renowned characters, and glorious works
Of high invention and of wond'rous art

Which not the ravages of time shall waste
Till he himself has run his long career;
Till all those glorious orbs of light on high,
The rolling wonders that surround the ball,
Drop from their spheres extinguish'd and consum'd;
When final ruin with her fiery car
Rides o'er creation, and all nature's works
Are lost in chaos and the womb of night.

2

TIMOTHY DWIGHT

The Conquest of Canäan; A Poem, in Eleven Books

1785

Timothy Dwight (1752–1817), a graduate of Yale College, chaplain of the Connecticut Continental Brigade, and a prominent Congregational minister in Connecticut, published his epic poem The Conquest of Canäan, *in Hartford in 1785. George Washington's name headed its list of subscribers. Heralded as America's first epic poem, the book was widely excerpted in American periodicals but coolly received by American readers.*

To his Excellency
GEORGE WASHINGTON, Esquire,
Commander in Chief of the American Armies.
The Saviour of his Country,
The Supporter of Freedom,
And the Benefactor of Mankind;
This Poem is inscribed,
With the highest Respect for his Character,
The most ardent Wishes for his Happiness,

Timothy Dwight, *The Conquest of Canäan; A Poem, in Eleven Books* (Hartford, 1785), dedication, v–vii, 296, 299–305.

And the most grateful Sense of the Blessings,
Secured, by his generous Efforts,
To the United States of North America.

> By his most humble,
> and most obedient Servant,
> TIMOTHY DWIGHT.

PREFACE

As this Poem is the first of the kind, which has been published in this country, the writer begs leave to introduce it with several observations, which that circumstance alone may perhaps render necessary. . . .

It may perhaps be thought the result of inattention or ignorance, that he chose a subject, in which his countrymen had no national interest. But he remarked, that the Iliad and Eneid were as agreeable to modern nations, as to the Greeks and Romans. The reason he supposed to be obvious—the subjects of those poems furnish the fairest opportunities of exhibiting the agreeable, the novel, the moral, the pathetic, and the sublime. If he is not deceived, the subject he has chosen possesses, in a degree, the same advantages.

It will be observed that he has introduced some new words, and annexed to some old ones, a new signification. This liberty, allowed to others, he hopes will not be refused to him: especially as from this source the copiousness and refinement of language have been principally derived.

That he wishes to please he frankly confesses. If he fails in the design, it will be a satisfaction that he shall have injured no person but himself. As the poem is uniformly friendly to delicacy, and virtue, he hopes his countrymen will so far regard him with candour, as not to impute it to him as a fault, that he has endeavoured to please them, and has thrown in his mite, for the advancement of the refined arts, on this side of the Atlantic.

GREENFIELD, IN CONNECTICUT,
   MARCH 1, 1785.

BOOK X

The Vision ceas'd. At once the forest fled,
At once an unknown region round them spread,
Like the still sabbath's dawning light serene,

And fair as blissful Eden's living green.
High on a hill they stood, whose cloudy brow
Look'd o'er th' illimitable world below.
In shining verdure eastern realms withdrew,
And hills and plains, immingling, fill'd the view:
From southern forests rose melodious sounds;
Tall, northern mountains stretch'd cerulean bounds;
West, all was sea; blue skies, with peaceful reign,
Serene roll'd round th' interminable plain.
Then thus the Power. To thee, bless'd man, 'tis given,
To know the thoughts of all-considering Heaven:
Scenes form'd eternal in th' unmeasur'd Mind,
In yon bright realms, for Abraham's race design'd,
While the great promise stands in heaven secure,
Or earth, or seas, or skies or stars endure.

. . .

Far o'er yon azure main thy view extend,*
Where seas, and skies, in blue confusion blend,
Lo, there a mighty realm, by heaven design'd
The last retreat for poor, oppress'd mankind!
Form'd with that pomp, which marks the hand divine,
And clothes yon vault, where worlds unnumber'd shine,
Here spacious plains in solemn grandeur spread;
Here cloudy forests cast eternal shade:
Rich vallies wind; the sky tall mountains brave,
And inland seas for commerce spread the wave;
With nobler floods, the sea-like rivers roll,
And fairer lustre purples round the pole.
Here, warm'd by happy suns, gay mines unfold
The useful iron, and the lasting gold;
Pure, changing gems in silence learn to glow,
And mock the splendors of the covenant bow:
On countless hills, by savage footsteps trod,
That smile to see the future harvest nod,
In glad succession, plants unnumber'd bloom,
And flowers unnumber'd breathe a rich perfume:
Hence life once more a length of days shall claim,
And health, reviving, light her purple flame.

*Vision of America.

Far from all realms this world imperial lies;
Seas roll between, and threatening storms arise;
Alike remov'd beyond Ambition's pale,
And the bold pinions of the venturous sail:
Till circling years the destin'd period bring,
And a new Moses lifts the daring wing,
Through trackless seas, an unknown flight explores,
And hails a new Canäan's promis'd shores.

On yon far strand, behold that little train*
Ascending, venturous, o'er th' unmeasur'd main.
No dangers fright: no ills the course delay:
'Tis virtue prompts, and GOD directs the way.
Speed, speed, ye sons of truth! let Heaven befriend,
Let angels waft you, and let peace attend!
O smile thou sky serene! ye storms retire!
And airs of Eden every sail inspire!
Swift o'er the main, behold the canvas fly,
And fade, and fade, beneath the farthest sky;
See verdant fields the changing waste unfold;
See sudden harvests dress the plains in gold;
In lofty walls the moving rocks ascend,
And dancing woods to spires and temples bend!

Meantime, expanding o'er earth's distant ends,
Lo, Slavery's gloom in sable pomp descends;
Far round each eastern clime her volumes roll,†
And pour, deep-shading, to the sadden'd pole.
How the world droops beneath the fearful blast;
The plains all wither'd, and the skies o'ercast!
From realm to realm extends the general groan;
The fainting body stupifies to stone;
Benumb'd, and fix'd, the palsied soul expires,
Blank'd all its views, and quench'd its living fires;
In clouds of boundless shade, the scenes decay;
Land after land departs, and nature fades away.

---

*Settlement of North America, by the English, for the enjoyment of Religion.
†Slavery of the eastern Continent.

In that dread hour, beneath auspicious skies,*
To nobler bliss yon western world shall rise.
Unlike all former realms, by war that stood,
And saw the guilty throne ascend in blood,
Here union'd Choice shall form a rule divine;
Here countless lands in one great system join;
The sway of Law unbroke, unrivall'd grow,
And bid her blessings every land o'erflow.

In fertile plains, behold the tree ascend,
Fair leaves unfold, and spreading branches bend!
The fierce, invading storm secure they brave,
And the strong influence of the creeping wave,
In heavenly gales with endless verdure rise,
Wave o'er broad fields, and fade in friendly skies.
There safe from driving rains, and battering hail,
And the keen fury of the wintery gale,
Fresh spring the plants; the flowery millions bloom,
All ether gladdening with a choice perfume;
Their hastening pinions birds unnumber'd spread,
And dance, and wanton, in th' aërial shade.

Here Empire's last, and brightest throne shall rise;
And Peace, and Right, and Freedom, greet the skies:
To morn's far realms her ships commercing fail,
Or lift their canvas to the evening gale;
In wisdom's walks, her sons ambitious soar,
Tread starry fields, and untried scenes explore.
And hark what strange, what solemn-breathing strain
Swells, wildly murmuring, o'er the far, far main!
Down time's long, lessening vale, the notes decay,
And, lost in distant ages, roll away.

When earth commenc'd, six morns of labour rose,†
Ere the calm Sabbath shed her soft repose.
Thus shall the world's great week direct its way,
And thousand circling suns complete the day.
Past were two days, ere beam'd the law divine;

*Freedom and glory of the North American States.
†The Jews have an ancient tradition of this nature.

Two days must roll, ere great Messiah shine;
Two changeful days, the Gospel's light shall rise:
Then sacred quiet hush the stormy skies;
O'er orient regions suns of toil shall roll,
Faint lustre dawn, and clouds obscure the pole:
But o'er yon favourite world, the Sabbath's morn,
Shall pour unbounded day, and with clear splendor burn.

Hence, o'er all lands shall sacred influence spread,
Warm frozen climes, and cheer the death-like shade;
To nature's bounds, reviving Freedom reign,
And Truth, and Virtue, light the world again.

3

JOEL BARLOW

The Vision of Columbus; A Poem in Nine Books
1787

Only a year after graduating from Yale College in 1777, Joel Barlow (1754–1812) began to write his epic poem, The Vision of Columbus. Finally completed and published in Hartford in 1787, it was the first American epic to focus on American nature, history, and futurity. Although the poem never achieved popularity, it was widely excerpted in American literary magazines and newspapers and went to a second edition in 1788. Barlow continued to revise his epic, even during his colorful career as a U.S. diplomat in Algiers and Europe.

INTRODUCTION

... The Author, at first, formed an idea of attempting a regular Epic Poem, on the discovery of America. But on examining the nature of that event, he found that the most brilliant subjects incident to such a

Joel Barlow, *The Vision of Columbus; A Poem in Nine Books* (Hartford, 1787), xxi, 158, 161–65, 198, 203–5.

plan would arise from the consequences of the discovery, and must be represented in vision. Indeed to have made it a patriotic Poem, by extending the subject to the settlement and revolutions of North America and their probable effect upon the future progress of society at large, would have protracted the vision to such a degree as to render it disproportionate to the rest of the work. To avoid an absurdity of this kind, which he supposed the critics would not pardon, he rejected the idea of a regular Epic form, and has confined his plan to the train of events which might be represented to the hero in vision. This form he considers as the best that the nature of the subject would admit; and the regularity of the parts will appear by observing, that there is a single poetical design constantly kept in view, which is to gratify and soothe the desponding mind of the hero: It being the greatest possible reward of his services, and the only one that his situation would permit him to enjoy, to convince him that his labours had not been bestowed in vain, and that he was the author of such extensive happiness to the human race. . . .

BOOK V

. . . Again the towns aspire, the cultured field
And blooming vale their copious treasures yield;
The grateful hind his cheerful labour proves,
And songs of triumph fill the warbling groves;
The conscious flocks, returning joys that share,
Spread thro' the midland, o'er the walks of war:
When, borne on eastern winds, dark vapors rise,
And sail and lengthen round the western skies;
Veil all the vision from his anxious sight,
And wrap the climes in universal night.

The hero grieved, and thus besought the Power:
Why sinks the scene? or must I view no more?
Must here the fame of that fair world descend?
And my brave children find so soon their end?
Where then the word of Heaven, Mine eyes should see
That half mankind should owe their bliss to me?

The Power replied: Ere long, in happier view,
The realms shall brighten, and thy joys renew.

The years advance, when round the thronging shore,
They rise confused to change the source of power;
When Albion's Prince,[3] that sway'd the happy land,
Shall stretch, to lawless rule, the sovereign hand;
To bind in slavery's chains the peaceful host,
Their rights unguarded and their charters lost.
Now raise thine eye; from this delusive claim,
What glorious deeds adorn their growing fame!

Columbus look'd; and still around them spread,
From south to north, the immeasurable shade;
At last, the central shadows burst away,
And rising regions open'd on the day.
He saw, once more, bright Del'ware's silver stream.
And Penn's throng'd city cast a cheerful gleam:
The dome of state, that met his eager eye,
Now heaved its arches in a loftier sky;
The bursting gates unfold; and lo, within,
A solemn train, in conscious glory, shine.
The well-known forms his eye had traced before,
In different realms along the extended shore;
Here, graced with nobler fame, and robed in state,
They look'd and moved magnificently great.

High on the foremost seat, in living light,
Majestic Randolph[4] caught the hero's sight:
Fair on his head, the civic crown was placed,
And the first dignity his sceptre graced.
He opes the cause, and points in prospect far,
Thro' all the toils that wait the impending war—
But, hapless sage, thy reign must soon be o'er,
To lend thy lustre and to shine no more.
So the bright morning star, from shades of even,
Leads up the dawn, and lights the front of heaven,
Points to the waking world the sun's broad way,
Then veils his own and shines above the day.
And see great Washington behind thee rise,

---

[3] *Albion's Prince:* the king of England
[4] *Peyton Randolph* (c. 1721–1775), the leading member of Virginia's most prominent family, was elected president of the First Continental Congress in 1774.

Thy following sun, to gild our morning skies;
O'er shadowy climes to pour the enlivening flame,
The charms of freedom and the fire of fame.
The ascending chief adorn'd his splendid seat,
Like Randolph, ensign'd with a crown of state;
Where the green patriot bay beheld, with pride,
The hero's laurel springing by its side;
His sword hung useless, on his graceful thigh,
On Britain still he cast a filial eye;
But sovereign fortitude his visage bore,
To meet their legions on the invaded shore.

Sage Franklin next arose, in awful mien,
And smiled, unruffled, o'er the approaching scene;
High on his locks of age a wreath was braced,
Palm of all arts, that e'er a mortal graced;
Beneath him lies the sceptre kings have borne,
And crowns and laurels from their temples torn.
Nash, Rutledge, Jefferson, in council great,
And Jay and Laurens oped the rolls of fate;
The Livingstons, fair Freedom's generous band,
The Lees, the Houstons, fathers of the land,
O'er climes and kingdoms turn'd their ardent eyes,
Bade all the oppress'd to speedy vengeance rise;
All powers of state, in their extended plan,
Rise from consent to shield the rights of man.
Bold Wolcott urged the all-important cause;
With steady hand the solemn scene he draws;
Undaunted firmness with his wisdom join'd,
Nor kings nor worlds could warp his stedfast mind.[5]

. . .

Adams, enraged, a broken charter bore,
And lawless acts of ministerial power;
Some injured right, in each loose leaf appears,
A king in terrors and a land in tears;
From all the guileful plots the veil he drew,
With eye retortive look'd creation thro',

[5]American Revolutionary leaders Benjamin Franklin, Abner Nash, John Rutledge, Thomas Jefferson, John Jay, Henry Laurens, the Livingston family of New York, the Lees of Virginia, the Houstons of New Jersey, Oliver Wolcott, John Adams

Oped the wide range of nature's boundless plan,
Traced all the steps of liberty and man;
Crouds [crowds] rose to vengeance while his accents rung,
And Independence thunder'd from his tongue.

BOOK VII

... Silent he gazed; when thus the guardian Power—
These works of peace awhile adorn the shore;
But other joys and deeds of lasting praise
Shall crown their labours and thy rapture raise.
Each orient realm, the former pride of earth,
Where men and science drew their ancient birth,
Shall soon behold, on this enlighten'd coast,
Their fame transcended and their glory lost.
That train of arts, that graced mankind before,
Warm'd the glad sage or taught the Muse to soar,
Here with superior sway their progress trace,
And aid the triumphs of thy filial race;
While rising crouds, with genius unconfined,
Through deep inventions lead the astonish'd mind,
Wide o'er the world their name unrivall'd raise,
And bind their temples with immortal bays.

In youthful minds to wake the ardent flame,
To nurse the arts, and point the paths of fame,
Behold their liberal sires, with guardian care,
Thro' all the realms their seats of science rear.
Great-without pomp the modest mansions rise;
Harvard and Yale and Princeton greet the skies;
Penn's ample walls o'er Del'ware's margin bend,
On James's bank the royal spires ascend,
Thy turrets, York, Columbia's walks command,
Bosom'd in groves, see growing Dartmouth stand;
While, o'er the realm reflecting solar fires,
On yon tall hill Rhode-Island's seat aspires.

O'er all the shore, with sails and cities gay,
And where rude hamlets stretch their inland sway,
With humbler walls unnumber'd schools arise,

And youths unnumber'd sieze the solid prize.
In no blest land has Science rear'd her fane,[6]
And fix'd so firm her wide-extended reign;
Each rustic here, that turns the furrow'd soil,
The maid, the youth, that ply mechanic toil,
In freedom nursed, in useful arts inured,
Know their just claims, and see their rights secured.

And lo, descending from the seats of art,
The growing throngs for active scenes depart;
In various garbs they tread the welcome land,
Swords at their side or sceptres in their hand,
With healing powers bid dire diseases cease,
Or sound the tidings of eternal peace.

In no blest land has fair Religion shone,
And fix'd so firm her everlasting throne.
Where, o'er the realms those spacious temples shine,
Frequent and full the throng'd assemblies join;
There, fired with virtue's animating flame,
The sacred task unnumber'd sages claim;
The task, for angels great; in early youth,
To lead whole nations in the walks of truth,
Shed the bright beams of knowledge on the mind,
For social compact harmonize mankind,
To life, to happiness, to joys above,
The soften'd soul with ardent zeal to move;
For this the voice of Heaven, in early years,
Tuned the glad songs of life-inspiring seers,
For this consenting seraphs leave the skies,
The God compassionates,[7] the Saviour dies.

Tho' different faiths their various orders show,
That seem discordant to the train below;
Yet one blest cause, one universal flame,
Wakes all their joys and centres every aim;
They tread the same bright steps, and smoothe the road,
Lights of the world and messengers of God.
So the galaxy broad o'er heaven displays

[6]*fane:* temple
[7]*compassionates:* treats with compassion

Of various stars the same unbounded blaze;
Where great and small their mingling rays unite,
And earth and skies repay the friendly light.

4

BENJAMIN FRANKLIN

A Scheme for a New Alphabet and Reformed Mode of Spelling

1768

As early as 1768, Benjamin Franklin (1706–1790) developed A Scheme for a New Alphabet and Reformed Mode of Spelling *to simplify and regularize English spelling. The radical scheme, which introduced many new characters to convey pure sounds, attracted little support over the years. Characteristically, Franklin withdrew it from public consideration until Noah Webster expressed interest in orthographic reform in 1786. The eighty-year-old Franklin gave Webster his enthusiastic support, and Webster frequently cited Franklin's* Scheme *in his own lectures and publications.*

Table of the Reformed Alphabet

CHARACTERS	SOUNDED RESPECTIVELY, AS IN THE WORDS IN THE COLUMN BELOW
o	Old.
ɑ	John, folly; awl, ball.
a	Man, can.
e	Men, lend, name, lane.
i	Did, sin, deed, seen.
u	Tool, fool, rule.
y	um, un; as in umbrage, unto, &c., and as in *er.*
h	Hunter, happy, high.

Benjamin Franklin, *A Scheme for a New Alphabet and Reformed Mode of Spelling* (1768), in *The Writings of Benjamin Franklin,* ed. Albert Henry Smyth (New York, 1907), 5:170–71, 174.

CHARACTERS	SOUNDED RESPECTIVELY, AS IN THE WORDS IN THE COLUMN BELOW
g	Give, gather.
k	Keep, kick.
ƞ	(sh) Ship, wish.
ɳ	(ng) ing, repeating, among.
n	End.
r	Art.
t	Teeth.
d	Deed.
l	Ell, tell.
s	Essence.
z	(ez) Wages.
ɳ	(th) Think.
ɳ	(dh) Thy.
f	Effect.
v	Ever.
b	Bees.
p	Peep.
m	Ember.

NAMES OF LETTERS AS EXPRESSED IN THE REFORMED SOUNDS AND CHARACTERS	MANNER OF PRONOUNCING THE SOUNDS [EXCERPT]
o	The first VOWEL naturally, and deepest sound; requires only to open the mouth, and breathe through it.
ɑ	The next requiring the mouth opened a little more, or hollower.
a	The next, a little more.
e	The next rquires the *tongue* to be a little more elevated.
i	The next still more.
u	The next requires the *lips* to be gathered up, leaving a small opening.
ɥ	The next a very short vowel, the sound of which we should express in our present letters thus, *uh;* a short, and not very strong *aspiration.*
huh	A stronger or more forcible aspiration.
gi	The first CONSONANT; being formed by the *root of the tongue;* this is the present hard *g*.
ki	A kindred sound; a little more acute; to be used instead of the hard *c*.

NAMES OF LETTERS AS EXPRESSED IN THE REFORMED SOUNDS AND CHARACTERS	MANNER OF PRONOUNCING THE SOUNDS [EXCERPT]
ish	A new letter wanted in our language; our *sh,* separately taken, not being proper elements of the sound.
ing	A new letter wanted for the same reason: —These are formed *back in the mouth.*
en	Formed *more forward* in the mouth; the *tip of the tongue* to the *roof* of the mouth.
r	The same; the tip of the tongue a little loose or separate from the roof of the mouth, and vibrating.
ti	The tip of the tongue more forward; touching, and then leaving, the roof.
di	The same; touching a little fuller.
el	The same; touching just about the *gums* of the *upper teeth.*
es	This sound is formed by the breath passing *between* the moist end of the *tongue* and the *upper teeth.*
ez	The same, a little denser and duller.
eh	The tongue under, and a little *behind,* the upper teeth; touching them, but so as to let the breath pass between.
eh̄	The same; a little fuller.
ef	Formed by the *lower lip* against the upper teeth.
ev	The same; fuller and duller.
b	The *lips full together,* and *opened* as the air passes out.
pi	The same; but a thinner sound.
em	The *closing* of the lips, while the *e* [here annexed] is sounding.

EXAMPLES [OF FRANKLIN'S REFORMED SPELLING]

So huen sym endfiel, byi divyin kamand,
Uih ryiziŋ tempests fieeks e gilti land,
(Sytfi az av leet or peel Britania past,)
Kalm and siriin hi dryivs hi fiuriys blast;
And, pliiz'd h' almyitis ardyrs tu pyrfarm,
Ryids in hi huyrluind and dyirekts hi starm.

So hi piur limpid striim, huen faul uih steens
av ryfiiŋ tarents and disendiŋ reens,
Uyrks itself kluiir; and az it ryns rifyins;
Til byi digriis, hi flotiŋ miryr fiyins,
Riflekts iitfi flaur hat an its bardyr groz,
And e nu hev'n in its feer byzym fioz.

NOAH WEBSTER

Dissertations on the English Language: with Notes, Historical and Critical

1789

Noah Webster (1758–1843), a graduate of Yale College and author of America's best-selling spelling book and first dictionary, published his Dissertations on the English Language *in Boston in 1789 to promote the cause of orthographic reform and the creation of an American language. Dedicated to Benjamin Franklin, the book was based on a series of public lectures that Webster had delivered in America's major towns. Both the lectures and the book were more popular in Webster's native New England than in the rest of the country.*

APPENDIX

AN ESSAY ON THE NECESSITY, ADVANTAGES AND PRACTICABILITY
OF REFORMING THE MODE OF SPELLING, AND OF RENDERING THE
ORTHOGRAPHY OF WORDS CORRESPONDENT TO THE PRONUNCIATION.

It has been observed by all writers on the English language, that the orthography or spelling of words is very irregular; the same letters often representing different sounds, and the same sounds often expressed by different letters. For this irregularity, two principal causes may be assigned:

1. The changes to which the pronunciation of a language is liable, from the progress of science and civilization.
2. The mixture of different languages occasioned by revolutions in England, or by a predilection of the learned, for words of foreign growth and ancient origin.

To the first cause, may be ascribed the difference between the spelling and pronunciation of Saxon words. The northern nations of

Noah Webster, *Dissertations on the English Language: with Notes, Historical and Critical* (Boston, 1789), 391–98, 405–6.

Europe originally spoke much in gutturals. This is evident from the number of aspirates and guttural letters, which still remain in the orthography of words derived from those nations; and from the modern pronunciation of the collateral branches of the Teutonic, the Dutch, Scotch and German. Thus *k* and *n* was once pronounced; as in *knave, know;* the *gh* in *might, though, daughter,* and other similar words; the *g* in *reign, feign,* &c.

But as savages proceed in forming languages, they lose the guttural sounds, in some measure, and adopt the use of labials, and the more open vowels. The ease of speaking facilitates this progress, and the pronunciation of words is softened, in proportion to a national refinement of manners. This will account for the difference between the ancient and modern languages of France, Spain and Italy; and for the difference between the soft pronunciation of the present languages of those countries, and the more harsh and guttural pronunciation of the northern inhabitants of Europe.

In this progress, the English have lost the sounds of most of the guttural letters. The *k* before *n* in *know,* the *g* in *reign,* and in many other words, are become mute in practice; and the *gh* is softened into the sound of *f,* as in *laughs,* or is silent, as in *brought.*

To this practice of softening the sounds of letters, or wholly suppressing those which are harsh and disagreeable, may be added a popular tendency to abbreviate words of common use. Thus *Southwark,* by a habit of quick pronunciation, is become *Suthark; Worcester* and *Leicester,* are become *Wooster* and *Lester; business, bizness; colonel, curnel; cannot, wilt not, cant, wont.** In this manner the final *e* is not heard in many modern words, in which it formerly made a syllable. The words *clothes, cares,* and most others of the same kind, were formerly pronounced in two syllables.[†]

... When words have been introduced from a foreign language into the English, they have generally retained the orthography of the original, however ill adapted to express the English pronunciation. Thus *fatigue, marine, chaise,* retain their French dress, while, to represent the true pronunciation in English, they should be spelt *fateeg, mareen, shaze.* Thus thro an ambition to exhibit the etymology of words, the English, in *Philip, physic, character, chorus,* and other Greek derivatives, preserve the representatives of the original Φ and Χ; yet these

---

Wont is strictly a contraction of *woll not,* as the word was anciently pronounced.
[†]"Ta-ke, ma-ke, o-ne, bo-ne, sto-ne, wil-le, &. dissyllaba olim fuerunt, quæ nunc habenter pro monosyllabis."—Wallis.

words are pronounced, and ought ever to have been spelt, *Fillip, fyzzic, or fizzic, karacter, korus.* *

But such is the state of our language. The pronunciation of the words which are strictly *English,* has been gradually changing for ages, and since the revival of science in Europe, the language has received a vast accession of words from other languages, many of which retain an orthography very ill suited to exhibit the true pronunciation.

The question now occurs; ought the Americans to retain these faults which produce innumerable inconveniencies in the acquisition and use of the language, or ought they at once to reform these abuses, and introduce order and regularity into the orthography of the AMERICAN TONGUE? ...

The principal alterations, necessary to render our orthography sufficiently regular and easy, are these:

1. The omission of all superfluous or silent letters; as *a* in *bread.* Thus *bread, head, give, breast, built, meant, realm, friend,* would be spelt, *bred, hed, giv, brest, bilt, ment, relm, frend.* Would this alteration produce any inconvenience, any embarrassment or expense? By no means. On the other hand, it would lessen the trouble of writing, and much more, of learning the language; it would reduce the true pronunciation to a certainty; and while it would assist foreigners and our own children in acquiring the language, it would render the pronunciation uniform, in different parts of the country, and almost prevent the possibility of change.

2. A substitution of a character that has a certain definite sound, for one that is more vague and indeterminate. Thus by putting *ee* instead of *ea* or *ie,* the words *mean, near, speak, grieve, zeal,* would become *meen, neer, speek, greev, zeel.* This alteration could not occasion a moments trouble; at the same time it would prevent a doubt respecting the pronunciation; whereas the *ea* and *ie* having different sounds, may give a learner much difficulty. Thus *greef* should be sub-

---

*The words *number, chamber,* and many other in English are from the French *nombre, chambre,* &c. Why was the spelling changed? or rather why is the spelling of *lustre, metre, theatre, not* changed? The cases are precisely similar. The Englishman who first wrote *number* for *nombre,* had no greater authority to make the change, than any modern writer has to spell *lustre, metre* in a similar manner, *luster, meter.* The change in the first instance was a valuable one; it conformed the spelling to the pronunciation, and I have taken the liberty, in all my writings, to pursue the principle in *luster, meter, miter, theater, sepulcher,* &c.

stituted for *grief, kee* for *key; beleev* for *believe; laf* for *laugh; dawter* for *daughter; plow* for *plough; tuf* for *tough; proov* for *prove; blud* for *blood;* and *draft* for *draught.* In this manner *ch* in Greek derivatives, should be changed into *k;* for the English *ch* has a soft sound, as in *cherish;* but *k* always a hard sound. Therefore *character, chorus, cholic, architecture,* should be written *karacter, korus, kolic, arkitecture;* and were they thus written, no person could mistake their true pronunciation.

Thus *ch* in French derivatives should be changed into *sh; machine, chaise, chevalier,* should be written *masheen, shaze, shevaleer;* and *pique, tour, oblique,* should be written *peek, toor, obleek.*

3. A trifling alteration in a character, or the addition of a point would distinguish different sounds, without the substitution of a new character. Thus a very small stroke across *th* would distinguish its two sounds. A point over a vowel, in this manner, *ȧ,* or *ȯ,* or *ī,* might answer all the purposes of different letters. And for the dipthong *ow,* let the two letters be united by a small stroke, or both engraven on the same piece of metal with the left hand line of the *w* united to the *o.*

These, with a few other inconsiderable alterations, would answer every purpose, and render the orthography sufficiently correct and regular.

The advantages to be derived from these alterations are numerous, great and permanent.

1. The simplicity of the orthography would facilitate the learning of the language. It is now the work of years for children to learn to spell; and after all, the business is rarely accomplished. A few men, who are bred to some business that requires constant exercise in writing, finally learn to spell most words without hesitation; but most people remain, all their lives, imperfect masters of spelling, and liable to make mistakes, whenever they take up a pen to write a short note. Nay, many people, even of education and fashion, never attempt to write a letter, without frequently consulting a dictionary.

But with the proposed orthography, a child would learn to spell, without trouble, in a very short time, and the orthography being very regular, he would ever afterwards find it difficult to make a mistake. It would, in that case, be as difficult to spell *wrong,* as it is now to spell *right.*

Besides this advantage, foreigners would be able to acquire the pronunciation of English, which is now so difficult and embarrassing, that they are either wholly discouraged on the first attempt, or obliged, after many years labor, to rest contented with an imperfect knowlege of the subject.

2. A correct orthography would render the pronunciation of the language, as uniform as the spelling in books. A general uniformity thro the United States, would be the event of such a reformation as I am here recommending. All persons, of every rank, would speak with some degree of precision and uniformity.* Such a uniformity in these states is very desireable; it would remove prejudice, and conciliate mutual affection and respect.

3. Such a reform would diminish the number of letters about one sixteenth or eighteenth. This would save a page in eighteen; and a saving of an eighteenth in the expense of books, is an advantage that should not be overlooked.

4. But a capital advantage of this reform in these states would be, that it would make a difference between the English orthography and the American. This will startle those who have not attended to the subject; but I am confident that such an event is an object of vast political consequence. For,

The alteration, however small, would encourage the publication of books in our own country. It would render it, in some measure, necessary that all books should be printed in America. The English would never copy our orthography for their own use; and consequently the same impressions of books would not answer for both countries. The inhabitants of the present generation would read the English impressions; but posterity, being taught a different spelling, would prefer the American orthography.

Besides this, a *national language* is a band of *national union*. Every engine should be employed to render the people of this country *national;* to call their attachments home to their own country; and to inspire them with the pride of national character. However they may boast of Independence, and the freedom of their government, yet their *opinions* are not sufficiently independent; an astonishing respect for the arts and literature of their parent country, and a blind imitation of its manners, are still prevalent among the Americans. Thus an habit-

---

*I once heard Dr. Franklin remark, "that those people spell best, who do not know how to spell"; that is, they spell as their ears dictate, without being guided by rules, and thus fall into a regular orthography.

ual respect for another country, deserved indeed and once laudable, turns their attention from their own interests, and prevents their respecting themselves. . . .

Sensible I am how much easier it is to *propose* improvements, than to *introduce* them. Every thing *new* starts the idea of difficulty; and yet it is often mere novelty that excites the appearance; for on a slight examination of the proposal, the difficulty vanishes. When we firmly *believe* a scheme to be practicable, the work is *half* accomplished. We are more frequently deterred by fear from making an attack, than repulsed in the encounter. . . .

But America is in a situation the most favorable for great reformations; and the present time is, in a singular degree, auspicious. The minds of men in this country have been awakened. New scenes have been, for many years, presenting new occasions for exertion; unexpected distresses have called forth the powers of invention; and the application of new expedients has demanded every possible exercise of wisdom and talents. Attention is roused; the mind expanded; and the intellectual faculties invigorated. Here men are prepared to receive improvements, which would be rejected by nations, whose habits have not been shaken by similar events.

Now is the time, and *this* the country, in which we may expect success, in attempting changes favorable to language, science and government. Delay, in the plan here proposed, may be fatal; under a tranquil general government, the minds of men may again sink into indolence; a national acquiescence in error will follow; and posterity be doomed to struggle with difficulties, which time and accident will perpetually multiply.

Let us then seize the present moment, and establish a *national language,* as well as a national government. Let us remember that there is a certain respect due to the opinions of other nations. As an independent people, our reputation abroad demands that, in all things, we should be federal; be *national;* for if we do not respect *ourselves,* we may be assured that *other nations* will not respect us. In short, let it be impressed upon the mind of every American, that to neglect the means of commanding respect abroad, is treason against the character and dignity of a brave independent people. . . .

6

NOAH WEBSTER

A Collection of Essays and Fugitiv Writings

1790

In A Collection of Essays and Fugitiv Writings, *published in Boston in 1790, Noah Webster modeled and defended the orthographic reforms that he had proposed in his* Dissertations on the English Language *of 1789. His fellow intellectuals and the public alike greeted the book with disbelief and ridicule. Stung by the public response, Webster retreated to a less radical position in his subsequent textbooks and dictionary, although privately he clung to his desire for an American language.*

PREFACE

The following Collection consists of Essays and Fugitiv Peeces, ritten at various times, and on different occasions, az wil appeer by their dates and subjects. Many of them were dictated at the moment, by the impulse of impressions made by important political events, and abound with a correspondent warmth of expression. This freedom of language wil be excused by the frends of the revolution and of good guvernment, who wil recollect the sensations they hav experienced, amidst the anarky and distraction which succeeded the cloze of the war. On such occasions a riter wil naturally giv himself up to hiz feelings, and hiz manner of *riting* wil flow from hiz manner of *thinking*.

Most of thoze peeces, which hav appeered before in periodical papers and Magazeens, were published with fictitious signatures; for I very erly discuvered, that although the name of an old and respectable karacter givs credit and consequence to hiz ritings, yet the name of a yung man iz often prejudicial to hiz performances. By conceeling my name, the opinions of men hav been prezerved from an undu bias arizing from personal prejudices, the faults of the ritings hav been detected, and their merit in public estimation ascertained.

Noah Webster, *A Collection of Essays and Fugitiv Writings, on Moral, Historical, Political and Literary Subjects* (Boston, 1790), ix–xi.

During the course of ten or twelve yeers, I hav been laboring to correct popular errors, and to assist my yung brethren in the road to truth and virtue; my publications for theze purposes hav been numerous; much time haz been spent, which I do not regret, and much censure incurred, which my hart tells me I do not dezerv. The influence of a yung riter cannot be so powerful or extensiv az that of an established karacter; but I hav ever thot a man's usefulness depends more on *exertion* than on *talents*. I am attached to America by berth, education and habit; but abuv all, by a philosophical view of her situation, and the superior advantages she enjoys, for augmenting the sum of social happiness.

In the essays, ritten within the last yeer, a considerable change of spelling iz introduced by way of experiment. This liberty waz taken by the riters before the age of queen Elizabeth, and to this we are indeted for the preference of modern spelling, over that of Gower and Chaucer. The man who admits that the change of *housbonde, mynde, ygone, moneth* into *husband, mind, gone, month,* iz an improovment, must acknowledge also the riting of *helth, breth, rong, tung, munth,* to be an improovment. There iz no alternativ. Every possible reezon that could ever be offered for altering the spelling of words, stil exists in full force; and if a gradual reform should not be made in our language, it wil proov that we are less under the influence of reezon than our ancestors.

Educating American Citizens

7

BENJAMIN RUSH

Thoughts Upon the Mode of Education Proper in a Republic

1786

Benjamin Rush (1745–1813), a graduate of the College of New Jersey, statesman, physician, and professor of chemistry and medicine in Philadelphia, was also one of the early Republic's leading educational theorists and reformers. His influential Thoughts Upon the Mode of Education Proper in a Republic *was published in Philadelphia in 1786 and widely reproduced in periodicals around the country.*

The business of education has acquired a new complexion by the independence of our country. The form of government we have assumed has created a new class of duties to every American. It becomes us, therefore, to examine our former habits upon this subject, and in laying the foundations for nurseries of wise and good men, to adapt our modes of teaching to the peculiar form of our government.

The first remark that I shall make upon this subject is that an education in our own is to be preferred to an education in a foreign country. The principle of patriotism stands in need of the reinforcement of

Benjamin Rush, *A Plan for the Establishment of Public Schools and the Diffusion of Knowledge in Pennsylvania; to Which are Added, Thoughts Upon the Mode of Education Proper in a Republic* (Philadelphia, 1786), 13–36.

prejudice, and it is well known that our strongest prejudices in favor of our country are formed in the first one and twenty years of our lives. The policy of the Lacedamonians[1] is well worthy of our imitation. When Antipater[2] demanded fifty of their children as hostages for the fulfillment of a distant engagement, those wise republicans refused to comply with his demand, but readily offered him double the number of their adult citizens, whose habits and prejudices could not be shaken by residing in a foreign country. Passing by, in this place, the advantages to the community from the early attachment of youth to the laws and constitution of their country, I shall only remark that young men who have trodden the paths of science together, or have joined in the same sports, whether of swimming, skating, fishing, or hunting, generally feel, thro' life, such ties to each other as add greatly to the obligations of mutual benevolence.

I conceive the education of our youth in this country to be peculiarly necessary in Pennsylvania, while our citizens are composed of the natives of so many different kingdoms in Europe. Our Schools of learning, by producing one general and uniform system of education, will render the mass of the people more homogeneous and thereby fit them more easily for uniform and peaceable government.

I proceed, in the next place, to enquire what mode of education we shall adopt so as to secure to the state all the advantages that are to be derived from the proper instruction of youth; and here I beg leave to remark that the only foundation for a useful education in a republic is to be laid in RELIGION. Without this, there can be no virtue, and without virtue there can be no liberty, and liberty is the object and life of all republican governments.

Such is my veneration for every religion that reveals the attributes of the Deity, or a future state of rewards and punishments, that I had rather see the opinions of Confucius or Mohammed inculcated upon our youth, than see them grow up wholly devoid of a system of religious principles. But the religion I mean to recommend in this place is the religion of JESUS CHRIST. . . .

Next to the duty which young men owe to their Creator, I wish to see a SUPREME REGARD TO THEIR COUNTRY inculcated upon them. When the Duke of Sully became prime minister to Henry the IVth of France, the first thing he did, he tells us, "was to subdue and forget his own heart." The same duty is incumbent upon every citizen of

---

[1]*Lacedamonians:* Spartans; residents of the Greek city-state of Sparta
[2]*Antipater:* Macedonian general

a republic. Our country includes family, friends and property, and should be preferred to them all. Let our pupil be taught that he does not belong to himself, but that he is public property. Let him be taught to love his family, but let him be taught at the same time that he must forsake and even forget them when the welfare of his country requires it.

He must watch for the state as if its liberties depended upon his vigilance alone, but he must do this in such a manner as not to defraud his creditors or neglect his family. He must love private life, but he must decline no station, however public or responsible it may be, when called to it by the suffrages of his fellow citizens. He must love popularity, but he must despise it when set in competition with the dictates of his judgement or the real interest of his country. He must love character and have a due sense of injuries, but he must be taught to appeal only to the laws of the state, to defend the one and punish the other. He must love family honour, but he must be taught that neither the rank nor antiquity of his ancestors can command respect without personal merit. He must avoid neutrality in all questions that divide the state, but he must shun the rage and acrimony of party spirit. He must be taught to love his fellow creatures in every part of the world, but he must cherish with a more intense and peculiar affection, the citizens of Pennsylvania and of the United States.

I do not wish to see our youth educated with a single prejudice against any nation or country; but we impose a task upon human nature, repugnant alike to reason, revelation, and the ordinary dimensions of the human heart, when we require him to embrace with equal affection the whole family of mankind. He must be taught to amass wealth, but it must be only to encrease his power of contributing to the wants and demands of the state. He must be indulged occasionally in amusements, but he must be taught that study and business should be his principal pursuits in life. Above all he must love life and endeavour to acquire as many of its conveniences as possible by industry and œconomy, but he must be taught that this life "is not his own," when the safety of his country requires it. These are practicable lessons, and the history of the commonwealths of Greece and Rome show that human nature, without the aids of Christianity, has attained these degrees of perfection.

While we inculcate these republican duties upon our pupil, we must not neglect at the same time to inspire him with republican principles. He must be taught that there can be no durable liberty but in a republic, and that government, like all other sciences, is of a progressive nature. The chains which have bound this science in Europe are happily unloosed in America. *Here* it is open to investigation and improvement. While philosophy has protected us by its discoveries from a thousand natural evils, government has unhappily followed with an unequal pace. It would be to dishonour human genius only to name the many defects which still exist in the best systems of legislation. We daily see matter of a perishable nature rendered durable by certain chemical operations. In like manner, I conceive that it is possible to analyze and combine power in such a manner as not only to encrease the happiness, but to promote the duration of republican forms of government far beyond the terms limited for them by history or the common opinions of mankind.

To assist in rendering religious, moral and political instruction more effectual upon the minds of our youth, it will be necessary to subject their bodies to physical discipline. To obviate the inconveniences of their studious and sedentary mode of life, they should live upon a temperate diet, consisting chiefly of broths, milk and vegetables. The black broth of Sparta and the barley broth of Scotland have been alike celebrated for their beneficial effects upon the minds of young people. They should avoid tasting spirituous liquors. They should also be accustomed occasionally to work with their hands in the intervals of study and in the busy seasons of the year in the country. Moderate sleep, silence, occasional solitude, and cleanliness should be inculcated upon them, and the utmost advantage should be taken of a proper direction of those great principles in human conduct—sensibility, habit, imitation, and association.

The influence [of] these physical causes will be powerful upon the intellects as well as upon the principles and morals of young people.

To those who have studied human nature, it will not appear paradoxical to recommend in this essay a particular attention to vocal music. Its mechanical effects in civilizing the mind and thereby preparing it for the influence of religion and government, have been so often felt and recorded that it will be unnecessary to mention facts in favour of its usefulness in order to excite a proper attention to it.

In the education of youth, let the authority of our masters be as *absolute* as possible. The government of schools, like the government of private families, should be *arbitrary,* that it may not be *severe.* By this mode of education, we prepare our youth for the subordination of laws and thereby qualify them for becoming good citizens of the republic. I am satisfied that the most useful citizens have been formed from those youth who have never known or felt their own wills till they were one and twenty years of age, and I have often thought that society owes a great deal of its order and happiness to the deficiencies of parental government being supplied by those habits of obedience and subordination which are contracted at schools.

I cannot help bearing a testimony, in this place, against the custom which prevails in some parts of America (but which is daily falling into disuse in Europe) of crouding boys together under one roof for the purpose of education. The practice is the gloomy remains of monkish ignorance and is as unfavourable to the improvements of the mind in useful learning as monasteries are to the spirit of religion. I grant this mode of secluding boys from the intercourse of private families has a tendency to make them scholars, but our business is to make them men, citizens and christians. The vices of young people are generally learned from each other. The vices of adults seldom infect them. By separating them from each other, therefore, in their hours of relaxation from study, we secure their morals from a principal source of corruption, while we improve their manners by subjecting them to those restraints which the difference of age and sex naturally produce in private families.

I have hitherto said nothing of the AMUSEMENTS that are proper for young people in a republic. Those which promote health and good humour will have a happy effect upon morals and government. To encrease this influence, let the persons who direct these amusements be admitted into good company and subjected by that means to restraints in behavior and moral conduct. Taverns, which in most countries are exposed to riot and vice, in Connecticut are places of business and innocent pleasure, because the tavern-keepers in that country are generally men of sober and respectable characters.

The theatre will never be perfectly reformed till players are treated with the same respect as persons of other ornamental professions. It is to no purpose to attempt to write or preach down an amusement which seizes so forcibly upon all the powers of the mind. Let ministers preach *to* players instead of *against* them; let them open their churches and the ordinances of religion to them and their families, and, I am persuaded, we shall soon see such a reformation in the theatre as can never be effected by all the means that have hitherto been employed for that purpose. It is possible to render the stage, by these means, subsurvient to the purposes of virtue and even religion. Why should the minister of the gospel exclude the player from his visits or from his public or private instructions? The Author of Christianity knew no difference in the occupations of men. He eat [ate] and drank daily with publicans[3] and sinners.

From the observations that have been made it is plain that I consider it as possible to convert men into republican machines. This must be done if we expect them to perform their parts properly in the great machine of the government of the state. That republic is sophisticated with monarchy or aristocracy that does not revolve upon the wills of the people, and these must be fitted to each other by means of education before they can be made to produce regularity and unison in government.

Having pointed out those general principles which should be inculcated alike in all the schools of the state, I proceed now to make a few remarks upon the method of conducting what is commonly called a liberal or learned education in a republic.

I shall begin this part of my subject by bearing a testimony against the common practice of attempting to teach boys the learned languages and the arts and sciences too early in life. The first twelve years of life are barely sufficient to instruct a boy in reading, writing and arithmetic. With these, he may be taught those modern languages which are necessary for him to *speak*. The state of the memory, in early life, is favourable to the acquisition of languages, especially when they are conveyed to the mind through the ear. It is, moreover, in early life only that the organs of speech yield in such a manner as to favour the just pronunciation of foreign languages.

I do not wish the LEARNED OR DEAD LANGUAGES, as they are commonly called, to be reduced below their present just rank in the universities of Europe, especially as I consider an acquaintance with them as the best foundation for a correct and extensive knowledge of the language of our country. Too much pains cannot be taken to teach our youth to read and write our American language with propriety and elegance. The study of the Greek language constituted a material part of the literature of the Athenians, hence the sublimity, purity and immortality of so many of their writings. The advantages of a perfect knowledge of our language to young men intended for the professions of law, physic or divinity are too obvious to be mentioned, but in a state which boasts of the first commercial city in America,[4] I wish to see it

---

[3] *publicans:* tax collectors
[4] Pennsylvania. Philadelphia was overtaken by New York City as America's leading commercial center in the second decade of the nineteenth century.

cultivated by young men who are intended for the compting house,[5] for many such, I hope, will be educated in our colleges. The time is past when an academical education was thought to be unnecessary to qualify a young man for merchandize. I conceive no profession is capable of receiving more embellishments from it.

Connected with the study of our own language is the study of ELOQUENCE. It is well known how great a part it constituted of the Roman education. It is the first accomplishment in a republic and often sets the whole machine of government in motion. Let our youth, therefore, be instructed in this art. We do not extol it too highly when we attribute as much to the power of eloquence as to the sword in bringing about the American revolution.

With the usual arts and sciences that are taught in our American colleges, I wish to see a regular course of lectures given upon HISTORY and CHRONOLOGY. The science of government, whether it relates to constitutions or laws, can only be advanced by a careful selection of facts, and these are to be found chiefly in history. Above all, let our youth be instructed in the history of the ancient republics, and the progress of liberty and tyranny in the different states of Europe.

I wish likewise to see the numerous facts that relate to the origin and present state of COMMERCE, together with the nature and principles of MONEY, reduced to such a system as to be intelligible and agreeable to a young man. If we consider the commerce of our metropolis only as the avenue of the wealth of the state, the study of it merits a place in a young man's education; but, I consider commerce in a much higher light when I recommend the study of it in republican seminaries. I view it as the best security against the influence of hereditary monopolies of land, and, therefore, the surest protection against aristocracy. I consider its effects as next to those of religion in humanizing mankind, and, lastly, I view it as the means of uniting the different nations of the world together by the ties of mutual wants and obligations.

CHEMISTRY, by unfolding to us the effects of heat and mixture, enlarges our acquaintance with the wonders of nature and the mysteries of art; hence it has become in most of the universities of Europe a necessary branch of a gentleman's education. In a young country, where improvements in agriculture and manufactures are so much to be desired, the cultivation of this science, which explains the

---

[5]*compting house:* counting house, business

principles of both of them, should be considered as an object of the utmost importance.

In a state where every citizen is liable to be a soldier and a legislator, it will be necessary to have some regular instruction given upon the ART OF WAR and upon PRACTICAL LEGISLATION. These branches of knowledge are of too much importance in a republic to be trusted to solitary study or to a fortuitous acquaintance with books. Let mathematical learning, therefore, be carefully applied in our colleges to gunnery and fortification, and let philosophy be applied to the history of those compositions which have been made use of for the terrible purposes of destroying human life. These branches of knowledge will be indispensably necessary in our republic, if unfortunately war should continue hereafter to be the unchristian mode of arbitrating disputes between Christian nations.

Again, let our youth be instructed in all the means of promoting national prosperity and independence, whether they relate to improvements in agriculture, manufactures, or inland navigation. Let him be instructed further in the general principles of legislation, whether they relate to revenue or to the preservation of life, liberty, or property. Let him be directed frequently to attend the courts of justice, where he will have the best opportunities of acquiring habits of arranging and comparing his ideas by observing the secretion of truth in the examination of witnesses and where he will hear the laws of the state explained, with all the advantages of that species of eloquence which belongs to the bar. Of so much importance do I conceive it to be to a young man to attend occasionally to the decisions of our courts of law, that I wish to see our colleges and academies established only in county towns.

But further, considering the nature of our connection with the United States, it will be necessary to make our pupil acquainted with all the prerogatives of the fœderal government. He must be instructed in the nature and variety of treaties. He must know the difference in the powers and duties of the several species of ambassadors. He must be taught wherein the obligations of individuals and of states are the same, and wherein they differ. In short, he must acquire a general knowledge of all those laws and forms which unite the sovereigns of the earth or separate them from each other.

I have only to add that it will be to no purpose to adopt this, or any other mode of education unless we make choice of suitable masters to carry our plans into execution. Let our teachers be distinguished for their abilities and knowledge. Let them be grave in their manners, gentle in their tempers, exemplary in their morals, and of sound principles in religion and government. Let us not leave their support to the precarious resources to be derived from their pupils, but let such funds be provided for our schools and colleges as will enable us to allow them liberal salaries. By these means we shall render the chairs,—the professorships and rectorships of our colleges and academies—objects of competition among learned men. By conferring upon our masters that independence which is the companion of competency, we shall, moreover, strengthen their authority over the youth committed to their care. Let us remember that a great part of the divines, lawyers, physicians, legislators, soldiers, generals, delegates, counsellors, and governors of the state will probably hereafter pass through their hands. How great then should be the wisdom, how honorable the rank, and how generous the reward of those men who are to form these necessary and leading members of the republic!

I beg pardon for having delayed so long, to say anything of the separate and peculiar mode of education proper for WOMEN in a republic. I am sensible that they must concur in all our plans of education for young men, or no laws will ever render them effectual. To qualify our women for this purpose, they should not only be instructed in the usual branches of female education, but they should be instructed in the principles of liberty and government; and the obligations of patriotism, should be inculcated upon them. The opinions and conduct of men are often regulated by the women in the most arduous enterprizes of life, and their approbation is frequently the principal reward of the hero's dangers and the patriot's toils. Besides, the *first* impressions upon the minds of children are generally derived from the women. Of how much consequence, therefore, is it in a republic that they should think justly upon the great subjects of liberty and government!

The complaints that have been made against religion, liberty, and learning have been made against each of them in a *separate* state. Perhaps like certain liquors they should only be used in a state of mixture. They mutually assist in correcting the abuses and in improving the good effects of each other. From the combined and reciprocal influence of religion, liberty, and learning upon the morals, manners and knowledge of individuals, of these upon government, and of government upon individuals, it is impossible to measure the degrees of happiness and perfection to which mankind may be raised. For my part, I can form no ideas of the golden age, so much celebrated by the poets, more delightful than the contemplation of that happiness which it is now in the power of the legislature of Pennsylvania to confer upon her citizens, by establishing proper modes and places of education in every part of the state.

The *present time* is peculiarly favourable to the establishment of these benevolent and necessary institutions in Pennsylvania. The minds of our people have not as yet lost the yeilding texture they acquired by the heat of the late Revolution. They will *now* receive more readily than five or even three years hence, new impressions and habits of all kinds. The spirit of liberty *now* pervades every part of the state. The influence of error and deception are *now* of short duration. Seven years hence, the affairs of our state may assume a new complexion. We may be rivetted to a criminal indifference for the safety and happiness of ourselves and our posterity. An aristocratic or democratic junto may arise that shall find its despotic views connected with the prevalence of ignorance and vice in the state; or a few artful pedagogues who consider learning as useful only in proportion as it favours their pride or avarice, may prevent all new literary establishments from taking place by raising a hue and cry against them, as the offspring of improper rivalship or the nurseries of party spirit.

But in vain shall we lavish pains and expence in establishing nurseries of virtue and knowledge in every part of the state, in vain shall we attempt to give the minds of our citizens a virtuous and uniform bias in early life, while the arms of our state are opened alike to receive into its bosom and to confer equal privileges upon the virtuous emigrant and the annual refuse of the jails of Britain, Ireland, and our sister states. Of the many criminals that have been executed within these seven years, four out of five of them have been foreigners who have arrived here during the war and since the peace. We are yet, perhaps, to see and deplore the tracks of the enormous vices and crimes these men have left behind them. Legislators of Pennsylvania!—Stewards of the justice and virtue of heaven!—Fathers of children who may be corrupted and disgraced by bad examples; say—can nothing be done to preserve our morals, manners, and government from the infection of European vices?

8

BENJAMIN RUSH

Plan of a Federal University

1788

Benjamin Rush's "Plan of a Federal University" was published in the Federal Gazette on October 29, 1788, and widely reprinted in the country's newspapers. James Madison of Virginia, James Wilson of Pennsylvania, and Charles Cotesworth Pinckney of South Carolina all expressed support for the plan, although it was never executed.

"Your government cannot be executed. It is too extensive for a republic. It is contrary to the habits of the people," say the enemies of the Constitution of the United States.—However opposite to the opinions and wishes of a majority of the citizens of the United States these declarations and predictions may be, they will certainly come to pass, unless the people are prepared for our new form of government by an education adapted to the new and peculiar situation of our country. To effect this great and necessary work, let one of the first acts of the new Congress be, to establish within the district to be allotted for them, a FEDERAL UNIVERSITY, into which the youth of the United States shall be received after they have finished their studies, and taken their degrees in the colleges of their respective states. In this University, let those branches of literature only be taught, which are calculated to prepare our youth for civil and public life. These branches should be taught by means of lectures, and the following arts and sciences should be the subjects of them.

1. The principles and forms of government, applied in a particular manner to the explanation of every part of the Constitution and laws of the United States, together with the laws of nature and nations, which last should include every thing that relates to peace, war, treaties, ambassadors, and the like.

2. History both ancient and modern, and chronology.

[Benjamin Rush], "Plan of a Federal University," *Federal Gazette and Philadelphia Evening Post*, October 29, 1788, 2–3.

3. Agriculture in all its numerous and extensive branches.
4. The principles and practice of manufactures.
5. The history, principles, objects and channels of commerce.
6. Those parts of mathematics which are necessary to the division of property, to finance, and to the principles and practice of war, for there is too much reason to fear that war will continue, for some time to come, to be the unchristian mode of deciding disputes between Christian nations.
7. Those parts of natural philosophy and chemistry, which admit of an application to agriculture, manufactures, commerce and war.
8. Natural history, which includes the history of animals, vegetables and fossils. To render instruction in these branches of science easy, it will be necessary to establish a museum, as also a garden, in which not only all the shrubs, &c. but all the forest trees of the United States should be cultivated. The great Linnæus of Upsal [Uppsala] enlarged the commerce of Sweden, by his discoveries in natural history. He once saved the Swedish navy by finding out the time in which a worm laid its eggs, and recommending the immersion of the timber, of which the ships were built, at that season wholly under water. So great were the services this illustrious naturalist rendered his country by the application of his knowledge to agriculture, manufactures and commerce, that the present king of Sweden pronounced an eulogium upon him from his throne, soon after his death.
9. Philology which should include, besides rhetoric and criticism, lectures upon the construction and pronunciation of the English language. Instruction in this branch of literature will become the more necessary in America, as our intercourse must soon cease with the bar, the stage and the pulpits of Great-Britain, from whence we received our knowledge of the pronunciation of the English language. Even modern English books should cease to be the models of stile in the United States. The present is the age of simplicity in writing in America. The turgid style of Johnson—the purple glare of Gibbon, and even the studied and thick set metaphors of Junius,[6] are all equally unnatural, and should not be admitted into our country. The cultivation and perfection of our language becomes a matter of consequence when viewed in another light. It will probably be spoken by more people in the course of two or three centuries, than ever spoke any one language at one time since the creation of the world. When

---

[6]Samuel Johnson (1709–1784), English lexicographer and author; Edward Gibbon (1737–1794), English historian; Franciscus Junius (1589–1677), English philologist.

we consider the influence which the prevalence of only *two* languages, viz. the English and the Spanish, in the extensive regions of North and South-America, will have upon manners, commerce, knowledge and civilization, scenes of human happiness and glory open before us, which elude from their magnitude the utmost grasp of the human understanding.

10. The German and French languages should be taught in this University. The many excellent books which are written in both these languages upon all subjects, more especially upon those which relate to the advancement of national improvements of all kinds, will render a knowledge of them an essential part of the education of a legislator of the United States.

11. All those athletic and manly exercises should likewise be taught in the University, which are calculated to impart health, strength, and elegance to the human body.

To render the instruction of our youth as easy and extensive as possible in several of the above mentioned branches of literature, let four young men of good education and active minds be sent abroad at the public expense, to collect and transmit to the professors of the said branches all the improvements that are daily made in Europe, in agriculture, manufactures and commerce, and in the art of war and practical government. This measure is rendered the more necessary from the distance of the United States from Europe, by which means the rays of knowledge strike the United States so partially, that they can be brought to a useful focus, only by employing suitable persons to collect and transmit them to our country. It is in this manner that the northern nations of Europe have imported so much knowledge from their southern neighbours, that the history of agriculture, manufactures, commerce, revenues and military arts of *one* of these nations will soon be alike applicable to all of them.

Besides sending four young men abroad to collect and transmit knowledge for the benefit of our country, *two* young men of suitable capacities should be employed at the public expense in exploring the vegetable, mineral and animal productions of our country, in procuring histories and samples of each of them, and in transmitting them to the professor of natural history. It is in consequence of the discoveries made by young gentlemen employed for these purposes, that Sweden, Denmark and Russia have extended their manufactures and commerce, so as to rival [those] in both the oldest nations in Europe.

Let the Congress allow a liberal salary to the Principal of this university. Let it be his business to govern the students, and to inspire them by his conversation, and by occasional public discourses, with federal and patriotic sentiments. Let this Principal be a man of extensive education, liberal manners and dignified deportment.

Let the Professors of each of the branches that have been mentioned, have a moderate salary of *150l.* or *200l.* [pound sterling] a year, and let them depend upon the number of their pupils to supply the deficiency of their maintenance from their salaries. Let each pupil pay for each course of lectures two or three guineas.

Let the degrees conferred in this university receive a new name, that shall designate the design of an education for civil and public life.

In thirty years after this university is established, let an act of Congress be passed to prevent any person being chosen or appointed into power or office, who has not taken a degree in the federal university. We require certain qualifications in lawyers, physicians and clergymen, before we commit our property, our lives or our souls to their care. We even refuse to commit the charge of a ship to a pilot, who cannot produce a certificate of his education and knowledge in his business. Why then should we commit our country, which includes liberty, property, life, wives and children, to men who cannot produce vouchers of their qualifications for the important trust? We are restrained from injuring ourselves by employing quacks in law; why should we not be restrained in like manner, by law, from employing quacks in government?

Should this plan of a federal university or one like it be adopted, then will begin the golden age of the United States. While the business of education in Europe consists in lectures upon the ruins of Palmyra and the antiquities of Herculaneum, or in disputes about Hebrew points, Greek particles, or the accent and quantity of the Roman language, the youth of America will be employed in acquiring those branches of knowledge which increase the conveniences of life, lessen human misery, improve our country, promote population, exalt the human understanding, and establish domestic, social and political happiness.

Let it not be said, "that this is not the *time* for such a literary and political establishment. Let us first restore public credit, by funding or paying our debts, let us regulate our militia, let us build a navy, and let us protect and extend our commerce. After this, we shall have leisure and money to establish a University for the purposes that have been mentioned." This is false reasoning. We shall never restore public credit, regulate our militia, build a navy, or revive our commerce, until we remove the ignorance and prejudices, and change the habits of our citizens, and this can never be done 'till we inspire them with federal principles, which can only be effected by our young men meeting and spending two or three years together in a national University, and afterwards disseminating their knowledge and principles through every county, township and village of the United States. 'Till this is done—Senators and Representatives of the United States, you will undertake to make bricks without straw. Your supposed union in Congress will be a rope of sand. The inhabitants of Massachusetts began the business of government by establishing the university of Cambridge [Harvard], and the wisest Kings in Europe have always found their literary institutions the surest means of establishing their power as well as of promoting the prosperity of their people.

These hints for establishing the Constitution and happiness of the United States upon a permanent foundation, are submitted to the friends of the federal government in each of the states, by a private

CITIZEN OF PENNSYLVANIA.

9

GEORGE WASHINGTON

Last Will and Testament

1799

Twice during his presidency, George Washington (1732–1799) formally (and futilely) asked Congress to establish a national university. His Last Will and Testament of 1799 bequeathed a small endowment for this purpose and explained why he thought such an institution necessary for the American Republic. Although Washington's disinterested gesture was much applauded, Congress never acted on his request.

In the name of God amen

I GEORGE WASHINGTON of Mount Vernon, a citizen of the United States, and lately President of the same, do make, orda[in] and

George Washington, *Last Will and Testament*, in *The Writings of George Washington, from the Original Manuscript Sources, 1745–1799*, ed. John C. Fitzpatrick (Washington, D.C., 1940), 37:275–81.

declare this Instrument; w[hic]h is written with my own hand [an]d every page thereof subscribed [wit]h my name, to be my last Will and [Tes]tament, revoking all others. . . .

Item   Whereas by a Law of the Commonwealth of Virginia, enacted in the year 1785, the Legislature thereof was pleased (as a an [sic] evidence of Its approbation of the services I had rendered the Public during the Revolution; and partly, I believe, in consideration of my having suggested the vast advantages which the Community would derive from the extension of its Inland Navigation, under Legislative patronage) to present me with one hundred shares of one hundred dollars each, in the incorporated company established for the purpose of extending the navigation of James River from tide water to the Mountains: and also with fifty shares of one hundred pounds Sterling each, in the Corporation of another company, likewise established for the similar purpose of opening the Navigation of the River Potomac from tide water to Fort Cumberland; the acceptance of which, although the offer was highly honorable, and grateful to my feelings, was refused, as inconsistent with a principle which I had adopted, and had never departed from, namely, not to receive pecuniary compensation for any services I could render my country in its arduous struggle with great Britain, for its Rights; and because I had evaded similar propositions from other States in the Union; adding to this refusal, however, an intimation that, if it should be the pleasure of the Legislature to permit me to appropriate the said shares to *public uses,* I would receive them on those terms with due sensibility; and this it having consented to, in flattering terms, as will appear by a subsequent Law, and sundry resolutions, in the most ample and honourable manner, I proceed after this recital, for the more correct understanding of the case, to declare:

That as it has always been a source of serious regret with me, to see the youth of these United States sent to foreign Countries for the purpose of Education, often before their minds were formed, or they had imbibed any adequate ideas of the happiness of their own; contracting, too frequently, not only habits of dissipation and extravagence, but principles unfriendly to Republican Governmt. and the true and genuine liberties of mankind; which, thereafter are rarely overcome. For these reasons, it has been my ardent wish to see a plan devised on a liberal scale which would have a tendency to sprd. systematic ideas through all parts of this rising Empire, thereby to do away [with] local attachments and State prejudices, as far as the nature of things would, or indeed ought to admit, from our National Councils. Looking anxiously forward to the accomplishment of so desirable an object as this is (in my estimation) my mind has not been able to contemplate any plan more likely to effect the measure than the establishment of a UNIVERSITY in a central part of the United States, to which the youth of fortune and talents from all parts thereof might be sent for the completion of their Education in all the branches of polite literature; in arts and Sciences, in acquiring knowledge in the principles of Politics and good Government; and (as a matter of infinite Importance in my judgment) by associating with each other, and forming friendships in Juvenile years, be enabled to free themselves in a proper degree from those local prejudices and habitual jealousies which have just been mentioned; and which, when carried to excess, are never failing sources of disquietude to the Public mind, and pregnant of mischievous consequences to this Country: Under these impressions, so fully dilated,

Item   I give and bequeath in perpetuity the fifty shares which I hold in the Potomac Company (under the aforesaid Acts of the Legislature of Virginia) towards the endowment of a UNIVERSITY to be established within the limits of the District of Columbia, under the auspices of the General Government, if that government should incline to extend a fostering hand towards it; and until such Seminary is established, and the funds arising on these shares shall be required for its support, my further WILL and desire is that the profit accruing therefrom shall, whenever the dividends are made, be laid out in purchasing Stock in the Bank of Columbia, or some other Bank, at the discretion of my Executors; or by the Treasurer of the United States for the time being under the direction of Congress; provided that Honourable body should Patronize the measure, and the Dividends proceeding from the purchase of such Stock is to be vested in more stock, and so on, until a sum adequate to the accomplishment of the object is obtained, of which I have not the smallest doubt, before many years passes away; even if no aid or encouraged is given by Legislative authority, or from any other source . . .

10

JUDITH SARGENT MURRAY

On the Equality of the Sexes
1790

One of New England's most prolific authors, Judith Sargent Murray (1751–1820) of Gloucester, Massachusetts, contributed regular essays to the Massachusetts Magazine, or, Monthly Museum of Knowledge and Rational Entertainment, *a Boston literary magazine published between 1789 and 1796. The second volume of the magazine (March/April 1790) held her essay "On the Equality of the Sexes," published under the pen name Constantia. All but completed in 1779, the essay exhibits Murray's early thoughts on gender and educational opportunity.*

That minds are not alike, full well I know,
This truth each day's experience will show;
To heights surprising some great spirits soar,
With inborn strength mysterious depths explore;
Their eager gaze surveys the path of light,
Confest it stood to Newton's piercing sight.

Deep science, like a bashful maid retires,
And but the *ardent* breast her worth inspires;
By perseverance the coy fair is won.
And Genius, led by Study, wears the crown.

But some there are who wish not to improve,
Who never can the path of knowledge love,
Whose souls almost with the dull body one,
With anxious care each mental pleasure shun;
Weak is the level'd, enervated mind,
And but while here to vegetate design'd.
The torpid spirit mingling with its clod,
Can scarcely boast its origin from God;

[Judith Sargent Murray], "On the Equality of the Sexes," *Massachusetts Magazine, or, Monthly Museum of Knowledge and Rational Entertainment,* March–? 1790, 132–35.

Stupidly dull—they move progressing on—
They eat, and drink, and all their work is done.
While others, emulous of sweet applause,
Industrious seek for each event a cause,
Tracing the hidden springs whence knowledge flows,
Which nature all in beauteous order shows.

Yet cannot I their sentiments imbibe,
Who this distinction to the sex ascribe,
As if a woman's form must needs enrol,
A weak, a servile, an inferiour soul;
And that the guise of man must still proclaim,
Greatness of mind, and him, to be the same:
Yet as the hours revolve fair proofs arise,
Which the bright wreath of growing fame supplies;
And in past times some men have *sunk* so *low,*
That female records nothing *less* can show.
But imbecility is still confin'd,
And by the lordly sex to us consign'd;
They rob us of the power t'improve,
And then declare we only trifles love;
Yet haste the era, when the world shall know,
That such distinctions only dwell below;
The soul unfetter'd, to no sex confin'd,
Was for the abodes of cloudless day design'd.

Mean time we emulate their manly fires,
Though erudition all their thoughts inspires,
Yet nature with *equality* imparts,
And *noble passions,* swell e'en *female hearts.*

Is it upon mature consideration we adopt the idea, that nature is thus partial in her distributions? Is it indeed a fact, that she hath yielded to one half of the human species so unquestionable a mental superiority? I know that to both sexes elevated understandings, and the reverse, are common. But, suffer me to ask, in what the minds of females are so notoriously deficient, or unequal. May not the intellectual powers be ranged under these four heads—imagination, reason, memory and judgment. The province of imagination hath long since been surrendered up to us, and we have been crowned undoubted sovereigns of the regions of fancy. Invention is perhaps the most arduous effort of the mind; this branch of imagination hath been particularly ceded to us, and we have been time out of mind invested with that creative faculty. Observe the variety of fashions (here I bar the contemptuous smile) which distinguish and adorn the female world; how continually are they changing, insomuch that they almost render the wise man's assertion problematical, and we are ready to say, *there is something new under the sun.* Now what a playfulness, what an exuberance of fancy, what strength of inventive imagination, cloth this continual variation discover? Again, it hath been observed, that if the turpitude of the conduct of our sex, hath been ever so enormous, so extremely ready are we, that the very first thought presents us with an apology, so plausible, as to produce our actions even in an amiable light. Another instance of our creative powers, is our talent for slander; how ingenious are we at inventive scandal? what a formidable story can we in a moment fabricate merely from the force of a prolifick imagination? how many reputations, in the fertile brain of a female, have been utterly despoiled? how industrious are we at improving a hint? suspicion how easily do we convert into conviction, and conviction, embellished by the power of eloquence, stalks abroad to the surprise and confusion of unsuspecting innocence. Perhaps it will be asked if I furnish these facts as instances of excellency in our sex. Certainly not; but as proofs of a creative faculty, of a lively imagination. Assuredly great activity of mind is thereby discovered, and was this activity properly directed, what beneficial effects would follow. Is the needle and kitchen sufficient to employ the operations of a soul thus organized? I should conceive not. Nay, it is a truth that those very departments leave the intelligent principle vacant, and at liberty for speculation. Are we deficient in reason? we can only reason from what we know, and if an opportunity of acquiring knowledge hath been denied us, the inferiority of our sex cannot fairly be deduced from thence. Memory, I believe, will be allowed us in common, since every one's experience must testify, that a loquacious old woman is as frequently met with, as a communicative old man; their subjects are alike drawn from the fund of other times, and the transactions of their youth, or of maturer life, entertain, or perhaps fatigue you, in the evening of their lives. "But our judgment is not so strong—we do not distinguish so well."—Yet it may be questioned, from what doth this superiority, in this determining faculty of the soul, proceed. May we not trace its source in the difference of education, and continued advantages? Will it be said that the judgment of a male of two years old, is more sage than that of a female's of the same age? I believe the reverse is generally observed to be true. But from that period what partiality! how is the one exalted, and the other depressed, by the contrary modes of education which are adopted! the one is taught to aspire, and the other is early confined and limitted. As their years increase, the sister must be wholly domesticated, while the brother is led by the hand through all the flowery paths of science. Grant that their minds are by nature equal, yet who shall wonder at the *apparent* superiority, if indeed custom becomes *second nature;* nay if it taketh place of nature, and that it doth the experience of each day will evince. At length arrived at womanhood, the uncultivated fair one feels a void, which the employments allotted her are by no means capable of filling. What can she do? to books she may not apply; or if she doth, *to those only of the novel kind,* lest she merit the appellation of a *learned lady,* and what ideas have been affixed to this term, the observation of many can testify. Fashion, scandal, and sometimes what is still more reprehensible, are then called in to her relief; and who can say to what lengths the liberties she takes may proceed. Meantime she herself is most unhappy; she feels the want of a cultivated mind. Is she single, she in vain seeks to fill up time from sexual employments or amusements. Is she united to a person whose soul nature made equal to her own, education hath set him so far above her, that in those entertainments which are productive of such rational felicity, she is not qualified to accompany him. She experiences a mortifying consciousness of inferiority, which embitters every enjoyment. Doth the person to whom her adverse fate hath consigned her, possess a mind incapable of improvement, she is equally wretched, in being so closely connected with an individual whom she cannot but despise. Now, was she permitted the same instructors as her brother, (with an eye however to their particular departments) for the employment of a rational mind an ample field would be opened. In astronomy she might catch a glimpse of the immensity of the Deity, and thence she would form amazing conceptions of the august and supreme Intelligence. In geography she would admire Jehovah in the midst of his benevolence; thus adapting this globe to the various wants and amusements of its inhabitants. In natural philosophy she would adore the infinite majesty of heaven, clothed in condescension; and as she traversed the reptile world, she would hail the goodness of a creating God. A mind, thus filled, would have little room for the trifles with which our sex are, with too much justice, accused of amusing themselves, and they would thus be rendered fit companions for those, who should one day wear them as their crown. Fashions, in their variety, would then give place to conjectures, which might perhaps conduce to the improvement of the literary world; and there would be no leisure for slander or detraction.

Reputation would not then be blasted, but serious speculations would occupy the lively imaginations of the sex. Unnecessary visits would be precluded, and that custom would only be indulged by way of relaxation, or to answer the demands of consanguinity and friendship. Females would become discreet, their judgments would be invigorated, and their partners for life being circumspectly chosen, an unhappy Hymen[7] would then be as rare, as is now the reverse.

Will it be urged that those acquirements would supersede our domestick duties? I answer that every requisite in female economy is easily attained; and, with truth I can add, that when once attained, they require no further *mental attention.* Nay, while we are pursuing the needle, or the superintendency of the family, I repeat, that our minds are at full liberty for reflection; that imagination may exert itself in full vigor; and that if a just foundation is early laid, our ideas will then be worthy of rational beings. If we were industrious we might easily find time to arrange them upon paper, or should avocations press too hard for such an indulgence, the hours allotted for conversation would at least become more refined and rational. Should it still be vociferated, "Your domestick employments are sufficient"—I would calmly ask, is it reasonable, that a candidate for immortality, for the joys of heaven, an intelligent being, who is to spend an eternity in contemplating the works of Deity, should at present be so degraded, as to be allowed no other ideas, than those which are suggested by the mechanism of a pudding, or the sewing the seams of a garment? Pity that all such censurers of female improvement do not go one step further, and deny their future existence; to be consistent they surely ought.

Yes, ye lordly, ye haughty sex, our souls are by nature *equal* to yours; the same breath of God animates, enlivens, and invigorates us; and that we are not fallen lower than yourselves, let those witness who have greatly towered above the various discouragements by which they have been so heavily oppressed; and though I am unacquainted with the list of celebrated characters on either side, yet from the observations I have made in the contracted circle in which I have moved, I dare confidently believe, that from the commencement of time to the present day, there hath been as many females, as males, who, by the *mere force of natural powers,* have merited the crown of applause; who, *thus unassisted,* have seized the wreath of fame. I know there are [those] who assert, that as the animal powers of the one sex are superiour, of course their mental faculties also must be stronger;

[7]*Hymen:* the Greek god of marriage

thus attributing strength of mind to the transient organization of this earth born tenement. But if this reasoning is just, man must be content to yield the palm [to] many of the brute creation, since by not a few of his brethren of the field, he is far surpassed in bodily strength. Moreover, was this argument admitted, it would prove too much, for occular demonstration evinceth, that there are many robust masculine ladies, and effeminate gentlemen. Yet I fancy that Mr. Pope,[8] though clogged with an enervated body, and distinguished by a diminutive stature, could nevertheless lay claim to greatness of soul; and perhaps there are many other instances which might be adduced to combat so unphilosophical an opinion. Do we not often see, that when the clay built tabernacle is well nigh dissolved, when it is just ready to mingle with the parent soil, the immortal inhabitant aspires to, and even attaineth heights the most sublime, and which were before wholly unexplored. Besides, were we to grant that animal strength proved any thing, taking into consideration the accustomed impartiality of nature, we should be induced to imagine, that she had invested the female mind with superiour strength as an equivalent for the bodily powers of man. But waving this however palpable advantage, for *equality only*, we wish to contend. . . .

[8]Alexander Pope, renowned eighteenth-century English poet

11

JUDITH SARGENT MURRAY

The Story of Margaretta

1798

In February 1792, Judith Sargent Murray launched a new series of essays in the Massachusetts Magazine *under the pen name Mr. Vigillius or "The Gleaner." Collected in three volumes and published in Boston in 1798,* The Gleaner *sold well and was well received in New England, although no second edition was printed. In an extended fiction called*

[Judith Sargent Murray], "The Story of Margaretta," in *The Gleaner* (Boston, 1798), 3 vols., I:1:66–76.

"The Story of Margaretta," Murray traces the education of Margaretta Melworth, a young orphan who is adopted by Mr. Vigillius and his wife, Mary. The maze of gender identities created as the views of a girl and a woman are filtered through the perspective of a man, who is himself a fictional creation of a woman, adds another layer of meaning to the story.

VII

Then smoothly spreads the retrospective scene,
When no gigantic errors intervene.

. . . When we returned home, we fitted up a little chamber, of which we constituted Margaretta the sole proprietor, my wife informing her that she should establish a post betwixt her apartment and her own, that if they chose, upon any occasion, to separate, they might with the greater convenience open a correspondence by letter. The rudiments of Margaretta's education had been attended to; in her plain work[9] she had made considerable proficiency; she could read the seventh, tenth, eleventh and twelfth chapters of Nehemiah, without much difficulty; and when her aunt was taken ill she was on the point of being put into joining-hand;[10] but Mary very soon sketched out for our charge rather an extensive plan of education; and as I was not entirely convinced of the inutility of her views, the natural indolence of my temper induced me to let the matter pass, without entering my caveat by way of stopping proceedings; and indeed, I think the propriety of circumscribing the education of a female, within such narrow bounds as are frequently assigned, is at least problematical. A celebrated writer, I really forget who, hath penned upon this subject a number of self-evident truths; and it is an incontrovertible fact, that to the matron is entrusted not only the care of her daughter, but also the forming the first and oftentimes the most important movements of that mind, which is to inform the future man; the early dawnings of reason she is appointed to watch, and from her are received the most indelible impressions of his life. Now, was she properly qualified, how enviable and how dignified would be her employment. The probability is, that the family of children, whom she directed, supposing them to possess

[9]*plain work:* sewing
[10]*joining-hand:* quilting

common capacities, being once initiated into the flowery paths of science, would seldom stop short of the desired goal. Fine writing, arithmetic, geography, astronomy, music, drawing; an attachment to all
these might be formed in infancy; the first principles of the fine arts
might be so accommodated, as to constitute the pastime of the child;
the seeds of knowledge might be implanted in the tender mind, and
even budding there, before the avocations of the father permitted him
to combine his efforts. Affection for the sweet preceptress, would
originate a strong predilection for instructions, that would with interesting tenderness be given, and that would be made to assume the
face of entertainment, and thus the young proficient would be, almost
imperceptibly, engaged in those walks, in which an advantageous perseverance might rationally be expected. A mother, who possesseth a
competent knowledge of the English and French tongues, and who is
properly assiduous about her children, I conceive, will find it little
more difficult to teach them to lisp in two languages, than in one; and
as the powers of the student advanceth, certain portions of the day
may be regularly appropriated to the converting in that language
which is not designed for the common intercourses of life. Letters, in
either tongue, to the parent, or fictitious characters, may be alter-
nately written, and thus an elegant knowledge of both may be gradually obtained. Learning, certainly, can never with propriety be
esteemed a burthen; and when the mind is judiciously balanced, it
renders the possessor not only more valuable, but also more amiable,
and more generally useful. Literary acquisitions cannot, unless the faculties of the mind are deranged, be lost: and while the goods of fortune may be whelmed beneath the contingencies of revolving time,
intellectual property still remains, and the mental funds can never be
exhausted. The accomplished, the liberally accomplished female, if
she is destined to move in the line of competency, will be regarded as
a pleasing and instructive companion; whatever she does will connect
an air of persuasive elevation; whatever she may be adventitiously
called, genuine dignity will be the accompaniment of her steps; she
will always be attended to with pleasure, and she cannot fail of being
distinguished; should she, in her career of life, be arrested by adverse
fortune, many resources of relief, of pleasure, and of emolument, open
themselves before her; and she is not *necessarily* condemned to laborious efforts, or to the drudgery of that unremitted sameness, which
the ro[u]tine of the needle presents.

But whatever may be the merits of the course which I am thus
apparently advocating, without stopping to examine the other side of

the question, I proceed to say, that the plan of education adopted for Margaretta was, as I have already hinted, sufficiently extensive, and that Mrs. Vigillius (to address my good wife, in her dignified character of governante,[11] with all possible respect) having instructed her pupil in the grand fundamental points of the philanthropic religion of Jesus, was never easy while any branch of improvement, which could by the most remote construction be deemed feminine, remained unessayed; and I must in justice declare, that the consequence, by producing Margaretta at the age of sixteen, a beautiful and accomplished girl, more than answered her most sanguine expectations.

Of needle work, in its varieties, my wife pronounced her a perfect mistress; her knowledge of the English, and French tongues, was fully adequate to her years, and her manner of reading had, for me, peculiar charms; her hand writing was neat and easy; she was a good accomptant,[12] a tolerable geographer and chronologist; she had skimmed the surface of astronomy and natural philosophy; had made good proficiency in her study of history and the poets; could sketch a landscape; could furnish, from her own fancy, patterns for the muslins which she wrought; could bear her part in a minuet and a cotillion, and was allowed to have an excellent hand upon the piano forte. We once entertained a design of debarring her the indulgence of novels; but those books, being in the hands of everyone, we conceived the accomplishment of our wishes in this respect, except we had bred her an absolute recluse, almost impracticable; and Mrs. Vigillius, therefore, thought it best to permit the use of every decent work, causing them to be read in her presence, hoping that she might, by her suggestions and observations, present an antidote to the poison, with which the pen of the novelist is too often fraught. The study of history was pursued, if I may so express myself, systematically: To the page of the historian one hour every day was regularly devoted; a second hour, Mary conversed with her adopted daughter upon the subject which a uniform course of reading had furnished; and a third hour Margaretta was directed to employ, in committing to paper such particular facts, remarks and consequences deduced therefrom, as had, during the hours appropriated to reading, and conversing, most strikingly impressed her mind; and by these means the leading features of history were indelibly imprinted thereon. Mrs. Vigillius also composed little geographical, historical, and chronological catechisms, or dialogues, the nature of which will be easily conceived; and she

[11]*governante:* governess
[12]*accomptant:* accountant

pronounced them of infinite advantage in the prosecution of her plan; she submitted likewise, at least once every week, to little voluntary absences, when my boy Plato, being constituted courier betwixt the apartments of my wife and daughter, an epistolary correspondence was carried on between them, from which more than one important benefit was derived; the penmanship of our charge was improved; the beautiful and elegant art of letter writing was by degrees acquired; and Margaretta was early accustomed to lay open her heart to her maternal friend.

Persons when holding the pen, generally express themselves more freely than when engaged in conversation; and if they have a perfect confidence in those whom they address, the probability is, that, unbosoming themselves, they will not fail to unveil the inmost recesses of their souls—thus was Margaretta properly and happily habituated to disclose, without a blush, each rising thought to her, on whom the care of preparing her for the great career of life had devolved.

No, Mr. Pedant, she was not unfitted for her proper sphere; and your stomach, however critical it may be, never digested finer puddings than those which I, with an uncommon zest, have partook, as knowing they were the composition of her fair hand—yes, in the receipts of cookery she is thoroughly versed; she is in every respect the complete housewife; and our linen never received so fine a gloss as when it was ironed and laid in order by Margaretta. Mrs. Vigillius was early taught the science of economy, and she took care to teach it to her daughter; and being more especially economical of time, she so arrangeth matters as never to appear embarrassed, or in a hurry, having always her hours of leisure, which she appropriates to the contingencies of the day. It is true, she does not often engage in visits of mere ceremony, seldom making one of any party, without some view either to her own emolument, or that of those about her; and with regard to dress, she spends but little time in assorting an article which is, it must be confessed, too generally a monopolizer of a blessing, that can hardly be too highly estimated. She doth not think it necessary to have her dishabille for the morning, her robe-de-chambre for noon, and her full trimmed polanee or trollopee,[13] for the evening. The morning, generally, except in cases of any particular emergency, presents her dressed for the day, and as she is always elegant, of course she can never be preposterous, extravagant or gaudy. It will be

---

[13]*polanee or trollopee:* polonaise, woman's dress consisting of a bodice with a skirt open over a petticoat

hardly necessary to add, that Miss Melworth was, and is, her exact copiest;[14] and indeed she is so warmly attached to my dear Mary, that I verily believe it would have been in her power to have initiated her into the devious paths of error; and this is saying a great deal of a mind which possesseth such innate goodness, as doth that which inhabits the gentle bosom of my Margaretta. . . .

But while we have been assiduously employed in cultivating the mind of Margaretta, we have been endeavouring to eradicate the seeds of that over-weening self conceit, which, while it would induce an ostentatious exhibition of those talents, natural, or adventitious, which she may possess—like a rampant weed would impede and overshadow the growth of every virtue. Against pride and affectation we have been careful to guard her, by constantly inculcating one grand truth; a truth, to the conviction of which every ingenuous mind must be ever open. Her person, the symmetry of her features, the rose and lily of her complexion, the *tout ensemble*[15] of her exterior, the harmony of her voice, &c. &c.—these are the endowments of nature— while the artificial accomplishments with which she is invested, resulting wholly from accident, and being altogether independent of her own arrangements, confer upon her no real or intrinsic merit.

We are daily assuring her, that every thing in future depends upon her own exertions, and that her character must be designated by that confutent [consistent] decency, that elegant propriety, and that dignified condescension, which are indeed truly estimable. We have apprized her, that in every stage of her journey through life, she will find friends—or a social intercourse with the circles in which she may be called to move—constituting one of her principal enjoyments, and that if she is not eager for admiration, if she avoids making a display of superior abilities, she will escape those shafts of envy which will otherwise be too surely aimed at her peace; and secure to herself the complacent feelings of those with whom she may be conversant.

Margaretta hath a becoming spirit, and dissimulation is a stranger to her heart; she is rather cheerful than gay; she never diverts herself with simplicity and ignorance; *double entendres* she detests; she is not an adept in the present fashionable mode of playing upon words, and she never descends to what is called jesting; she can deliver herself upon any subject, on which she ventures to speak, with great ease; but in large or mixed companies she engages in conversation with

---

[14]*copiest:* copyist; imitator
[15]*tout ensemble:* general effect

manifest reluctance; and I have heard her declare, that she hath frequently, when encircled by strangers, felt alarmed at the sound of her own voice; she never comments upon those blunders which are the result of a neglected education, nor will she lend her smiles to those who are thus employed; and she observes, that such kind of peccadillos have upon her no other effect, than to excite in her bosom the sensation of gratitude.

With the laws of custom, or fashion, she is thoroughly acquainted, and she consents to follow them as far as they square with the dictates of rectitude; but she never sacrifices to their documents either her humanity, or her convenience; she regards, as extremely venial, an ignorance of their despotic institutions; (indeed the multifarious requirements of mere ceremony, strike her in so trifling a point of view, that she conceives it rather a matter of course that they should sometimes be omitted) and she prefers plain manners to all the glitter of a studied or laboured address.

But it is against the unaccountable freaks of the capricious, that all the artillery of that humour, of which she possesses a natural fund, is levelled; frank and ingenuous herself, she laughs at the vagaries of the whimsical, and her heart is ever upon her lips; she reflects much, and her judgement is fashioned by reason; she cannot be seen without pleasure, nor heard without instruction. . . .

12

NOAH WEBSTER

The American Spelling Book

1789

Noah Webster's speller was published as the first part of his Grammatical Institute of the English Language *in 1783 and renamed* The American Spelling Book *in 1788. This first American textbook was also the most popular ever published: more than 50 impressions of the book appeared*

Noah Webster, *The American Spelling Book* (Boston, 1789), 54–57, 73–74, 119–23, 141.

before 1800, some of them 25,000-copy runs. By 1807, it was selling more than 200,000 copies a year; by 1829, at least 20 million copies of the book had been sold in the United States, and it had been read by almost every literate American.

Table XIII

Lessons of easy Words, to teach Children to read, and to know their Duty.

LESSON I

No man may put off the law of God:
My joy is in his law all the day.
O may I not go in the way of sin!
Let me not go in the way of ill men.

II

A bad man is a foe to the law:
It is his joy to do ill.
All men go out of the way.
Who can say he has no sin?

III

The way of man is ill.
My son, do as you are bid:
But if you are bid, do no ill.
See not my sin, and let me not go to the pit.

IV

Rest in the Lord, and mind his word.
My son, hold fast the law that is good.
You must not tell a lie, nor do hurt.
We must let no man hurt us.

V

Do well as you can, and do no harm.
Mark the man that doth well, and do so too.
Help such as want help, and be kind.
Let your sins past, put you in mind to mend.

VI

I will not walk with bad men, that I may not
   be cast off with them.
I will love the law and keep it.
I will walk with the just and do good.

VII

This life is not long; but the life to come has
   no end.
We must pray for them that hate us.
We must love them that love not us.
We must do as we like to be done to.

VIII

A bad life will make a bad end.
He must live well that will die well.
He doth live ill that doth not mend.
In time to come we must do no ill.

IX

No man can say that he has done no ill,
For all men have gone out of the way.
There is none that doth good; no not one.
If I have done harm, I must do it no more.

X

Sin will lead us to pain and wo[e].
Love that which is good, and shun vice.
Hate no man, but love both friends and foes.
A bad man can take no rest, day nor night.

XI

He who came to save us, will wash us from all sin; I will be glad in his
name.

   A good boy will do all that is just: he will flee from vice; he will do
good, and walk in the way of life.

   Love not the world, nor the things that are in the world; for they are
sin.

   I will not fear what flesh can do to me; for my trust is in him who
made the world.

   He is nigh to them that pray to him, and praise his name.

XII

Be a good child; mind your book; love your school, and strive to learn.

Tell no tales; call no ill names; you must not lie, nor swear, nor cheat, nor steal.

Play not with bad boys; use no ill words at play; spend your time well; live in peace, and shun all strife. This is the way to make good men love you, and save your soul from pain and woe.

XIII

A good child will not lie, swear, nor steal. — He will be good at home, and ask to read his book; when he gets up he will wash his hands and face clean; he will comb his hair, and make haste .o school; he will not play by the way as bad boys do.

XIV

When good boys and girls are at school, they will mind their books, and try to learn to spell and read well, and not play in the time of school.

When they are at church, they will sit, kneel, or stand still; and when they are at home, will read some good book, that God may bless them.

XV

As for those boys and girls that mind not their books, and love not the church and school, but play with such as tell tales, tell lies, curse, swear and steal, they will come to some bad end, and must be whipt till they mend their ways. . . .

The following are accented on the first and third syllables.

Ap per tain	con nois seur	em bra sure
ad ver tise	dis ap pear	ac qui esce
as cer tain	en ter tain	co a lesce
con tra vene	gaz et teer	male con tent
can non ade	deb o nair	coun ter mand

Table XXII

Words not exceeding three syllables, divided.

LESSON I

The wick-ed flee when no man pur-su-eth; but the right-e-ous are bold as a li-on.

Vir-tue ex-alt-eth a na-tion; but sin is a re-proach to a-ny peo-ple.

The law of the wise is a foun-tain of life to de-part from the snares of death.

Wealth got-ten by de-ceit, is soon wast-ed; but he that gath-er-eth by la-bor, shall in-crease in rich-es.

II

I-dle-ness will bring thee to pov-er-ty; but by in-dus-try and pru-dence thou shalt be fill-ed with bread.

Wealth mak-eth ma-ny friends; but the poor are for-got-ten by their neigh-bors.

A pru-dent man fore-seeth the e-vil, and hid-eth him-self; but the thought-less pass on and are pun-ish-ed.

III

Train up a child in the way he should go; and when he is old he will not de-part from it.

Where there is no wood the fire go-eth out, and where there is no tat-ler the strife ceas-eth.

A word fit-ly spok-en is like ap-ples of gold in pic-tures of sil-ver.

He that cov-er-eth his sins shall not pros-per, but he that con-fess-eth and for-sak-eth them shall find mer-cy.

IV

The rod and re-proof give wis-dom; but a child left to him-self bring-eth his pa-rents to shame.

Cor-rect thy son, and he will give thee rest; yea, he will give thee de-light to thy soul.

A man's pride shall bring him low; but hon-or shall up-hold the humble in spir-it.

The eye that mock-eth at his fath-er, and scorn-eth to o-bey his moth-er, the ra-vens of the val-ley shall pick it out, and the young ea-gles shall eat it.

V

By the bless-ing of the up-right, the cit-y is ex-alt-ed, but it is o-ver-thrown by the mouth of the wick-ed.

Where no coun-cil is, the peo-ple fall; but in the mul-ti-tude of coun-sel-lors there is safe-ty.

The wis-dom of the pru-dent is to un-der-stand his way, but the fol-ly of fools is de-ceit.

A wise man fear-eth and de-part-eth from e-vil, but the fool ra-geth and is con-fi-dent.

Be not hast-y in thy spir-it to be an-gry; for an-ger rest-eth in the bo-som of fools. . . .

Table LV

THE UNITED STATES OF AMERICA.

States	Capital Towns	Inhabitants
New-Hamp-shire	Ports-mouth	102,000
Mas-sa-chu-setts	Bos-ton	360,000
Rhode-Is-land	New-port	58,678
Con-nec-ti-cut	Hart-ford	209,150
New-York	New-York	238,897
New-Jer-sey	Tren-ton	138,000
Penn-syl-va-ni-a	Phi-la-del-phi-a	360,000
Del-a-ware	New-Cas-tle	37,000
Ma-ry-land	Bal-ti-more	253,000
Vir-gin-i-a	Rich-mond	567,614
North-Ca-ro-li-na	New-bern	270,000
South-Ca-ro-li-na	Charles-ton	180,000
Geor-gi-a	Sa-van-nah	98,000

NEW HAMPSHIRE

Counties	Counties
Rock-ing-ham	Chesh-ire
Hils-bo-rough	Graf-ton
Staf-ford	

MASSACHUSETTS

Counties	Capital Towns
Suf-folk	Bos-ton
Es-sex	Sa-lem
Mid-dle-sex	Cam-bridge
Hamp-shire	Spring-field
Plym-outh	Plym-outh
Barn-sta-ble	Barn-sta-ble
Bris-tol	Taun-ton

Counties	Capital Towns
York	York
Dukes-Coun-ty	Ed-gar-ton
Nan-tuck-et	Sher-burne
Worce-ster	Worce-ster
Cum-ber-land	Port-land
Lin-coln	Pow-nal-bo-rough
Berk-shire	Great-ba-ring-ton

RHODE ISLAND

Counties	Capital Towns
New-port	New-port
Wash-ing-ton	South-King-ston
Prov-i-dence	Prov-i-dence
Kent	East-Green-wich
Bris-tol	Bris-tol

CONNECTICUT

Counties	Capital Towns
Hart-ford	Hart-ford
New-Hav-en	New-Hav-en
New-Lon-don	New-Lon-don
Wind-ham	Wind-ham
Fair-field	Fair-field
Litch-field	Litch-field
Mid-dle-sex	Mid-dle-ton
Tol-and	Tol-and

NEW YORK

Counties	Capital Towns
New-York	The City
Rich-mond	Rich-mond
King's-coun-ty	Flat-bush
Queen's-coun-ty	Ja-mai-ca
Suf-folk	South-hold
Al-ba-ny	Al-ba-ny
West-Ches-ter	West-Ches-ter
Or-ange	Go-shen
Ul-ster	King-ston

Counties	Capital Towns
Duch-ess	Pough-keep-sie*
Mont-go-me-ry	John-stown
Wash-ing-ton	
Co-lum-bia	Clav-er-ak

*Pronounced Pokepse.

NEW JERSEY

Counties	Capital Towns
Ber-gen	Ber-gen
Mid-dle-sex	Am-boy
Es-sex	New-ark
Som-er-set	Prince-ton
Mon-mouth	Free-hold
Mor-ris	Mor-ris-ton
Cum-ber-land	Bridge-town
Sux-sex	New-ton
Bur-ling-ton	Bur-ling-ton
Glouce-ster	Had-don-field
Sa-lem	Sa-lem
Hun-ter-don	Tren-ton
Cape May	

PENNSYLVANIA

Counties	Capital Towns
Phi-la-del-phi-a	Phi-la-del-phi-a
Ches-ter	Ches-ter
Bucks	New-ton
Lan-cas-ter	Lan-cas-ter
York	York
Cum-ber-land	Car-lisle
Berks	Read-ing
North-amp-ton	Eas-ton
Bed-ford	Bed-ford
North-um-ber-land	Sun-bu-ry
West-more-land	Han-nah's-town
Wash-ing-ton	
Frank-lin	

PENNSYLVANIA (*continued*)

Counties	Capital Towns
Dau-phin	
Fay-ette	
Lu-zerne	Wilks-bar-ry

DELAWARE

Counties	Capital Towns
New-Cas-tle	New-Cas-tle
Kent	Do-ver
Sus-sex	Lew-is-town

MARYLAND

Counties. — Worcester, Somerset, Dorchester, Talbot, Queen Ann's, Kent, Caroline, Cecil, Washington, St. Mary's, Charles, Prince George, Montgomery, Frederick, Anne Arundle, Baltimore, Hartford, Calvert.

VIRGINIA

Counties. — Amherst, Henrico, Richmond, Ohio, Prince William, Charlotte, Williamsburgh, James City, Northumberland, Nansemond, Buckingham, King and Queen, Stafford, Mecklenburgh, Louisa, Dinwiddie, Essex, York, Prince Edward, Fairfax, Goochland, Culpepper, Cumberland, Brunswick, Fauquier, Middlesex, Warwick, Caroline, Southampton, Botetourt, Spotsylvania, Norfolk, Amelia, Elizabeth City, Shenandoah, Monongahela, Bedford, Yohogany, Rockingham, London, Frederick, Montgomery, Kentucky, Rockbridge, Northampton, Prince George, Hampshire, Augusta, Berkley, Greenbriar, Pittsylvania, Surry, Accomack, Westmoreland, Washington, Charles City, Isle of Wight, Hanover, King George, Gloucester, Fluvanna, Princess Ann, Albemarle, New Kent, Lunenburg, Sussex, Lancaster, Powhatan, Orange, Henry, Chesterfield.

NORTH CAROLINA

This State is divided into Seven Districts.

1. EDENTON. *Counties.* — Currituk, Camden, Pasquetank, Perquimins, Chawan, Gates, Hartford, Bertie, Tyrrel.
2. HALIFAX. *Counties.* — Northampton, Halifax, Franklin, Warren, Nash, Edgecomb, Martin.
3. NEWBERN. *Counties.* — Craven, Dobbs, Johnston, Pitt, Beaufort, Carteret, Jones, Wayne, Hyde.

4. WILMINGTON. *Counties.*—Onslow, New Hanover, Brunswick, Bladen, Duplin, Cumberland.
5. HILISBOROUGH. *Counties.*—Granville, Caswell, Orange, Wake, Randolph, Chatham.
6. MORGAN. *Counties.*—Burke, Wilkes, Rutherford, Washington, Sullivan, Lincoln.
7. SALISBURY. *Counties.*—Rowan, Anson, Meklenburgh, Guildford, Surry, Montgomery, Richmond

SOUTH CAROLINA

This State is divided into Seven Districts.
1. NINETY SIX. *Counties.*—Abbeville, Edgefield, Newbury, Laurena, Spartenburg, Union.
2. CAMDEN. *Counties.*—Clarendon, Fairfield, Claremont, Richland, Lancaster, York, Chester.
3. CHERAWS. *Counties.*—Marlborough, Chesterfield, Darlington.
4. GEORGETOWN. *Counties.*—Winyaw, Williamsburgh, Kingston, Liberty.
5. CHARLESTON. *Counties.*—Charleston, Washington, Marion, Berkely, Bartholomew, Colleton.
6. BEAUFORT. *Counties.*—Hilton, Lincoln, Shrewsbury, Granville.
7. ORANGEBURGH. *Counties.*—Lewisburgh, Lexington, Orange, Winter.

GEORGIA

Counties	*Capital Towns*
Chat-ham	Sa-van-nah
Ef-fing-ham	E-ben-e-zer
Burke	Waynes-bo-rough
Rich-mond	Au-gus-ta
Wilkes	Wash-ing-ton
Frank-lin	
Green	Greens-bo-rough
Wash-ing-ton	Gol-phin-ton
Lib-er-ty	Sun-bu-ry
Glynn	Bruns-wick
Cam-den	St. Pa-tricks
Bour-bon	on the Mississippi, unsettled.

FAMILIAR PHRASES, AND EASY DIALOGUES, FOR YOUNG BEGINNERS.

Lesson I.

Sir, your most humble servant.
I have the pleasure to be yours.
I hope you are very well.
I am very well, Sir, I thank you.
How do they do at your house?
They are all well.
And you, Madam, how do you do?
Pretty well.   Very well.
Is all your family well?
Perfectly well.
How does your father do, your mother and your sisters?
You do them much honor; they are all in good health.
I am glad of having the pleasure to see you in good health.
I am much obliged to you.
Now I think on it, how does your brother do?
Exceedingly well; or indifferently well.
Does your brother go to school?
Yes, Sir, and my sisters too.
What do they learn?
They learn writing and English grammar.
I hope they make good improvement of their time.
Their instructor tells us that they are diligent, and make good progress in their studies.
I am glad to hear it; I hope to have the pleasure of seeing them at the next holy-days.
Sir, they will be no less happy to see you.

Narrating Nationhood

13

DAVID RAMSAY

The History of the American Revolution

1789

David Ramsay (1749–1815), a physician, legislator, and historian, graduated from the College of New Jersey and studied medicine at the College of Philadelphia with Benjamin Rush. He accepted a position and settled permanently in Charleston, South Carolina, in 1774. His History of the American Revolution *was published in Philadelphia in 1789 and serialized in the* Columbian Magazine. *Six American editions of the book were published between 1789 and 1865, as well as German, Dutch, Irish, and two English and French editions.*

APPENDIX: NO. IV

The State of parties: the advantages and disadvantages of the Revolution: its influence on the minds and morals of the Citizens.

Previous to the American revolution, the inhabitants of the British Colonies were universally loyal. That three millions of such subjects should break through all former attachments, and unanimously adopt new ones, could not reasonably be expected. The revolution had its enemies, as well as its friends, in every period of the war. Country,

David Ramsay, *The History of the American Revolution* (Philadelphia, 1789), 2 vols., 2:310–17, 323–25, 340–41, 344–45, 354–56.

religion, local policy, as well as private views, operated in disposing the inhabitants to take different sides. The New-England provinces being mostly settled by one sort of people, were nearly of one sentiment. The influence of placemen[1] in Boston, together with the connections which they had formed by marriages, had attached sundry influential characters in that capital to the British interest, but these were but as the dust in the balance, when compared with the numerous independent Whig yeomanry of the country. The same and other causes produced a large number in New-York, who were attached to royal government. That city had long been head quarters of the British army in America, and many intermarriages, and other connections, had been made between British officers, and some of their first families. The practice of entailing estates had prevailed in New-York to a much greater extent, than in any of the other provinces. The governors thereof had long been in the habit of indulging their favourites with extravagant grants of land. This had introduced the distinction of landlord and tenant. There was therefore in New-York an aristocratic party, respectable for numbers, wealth, and influence, which had much to fear from independence. The city was also divided into parties by the influence of two ancient and numerous families, the Livingstones and Delanceys. These having been long accustomed to oppose each other at elections, could rarely be brought to unite in any political measures. In this controversy, one almost universally took part with America, the other with Great Britain.

The Irish in America, with a few exceptions, were attached to independence. They had fled from oppression in their native country, and could not brook the idea that it should follow them. Their national prepossessions in favour of liberty were strengthened by their religious opinions. They were Presbyterians, and people of that denomination, for reasons hereafter to be explained, were mostly Whigs. The Scotch on the other hand, though they had formerly sacrificed much to liberty in their own country, were generally disposed to support the claims of Great Britain. Their nation for some years past had experienced a large proportion of royal favour. A very absurd association was made by many, between the cause of John Wilkes and the cause of America. The former had rendered himself so universally odious to the Scotch, that many of them were prejudiced against a cause, which was so ridiculously, but generally associated, with that of a man who had grossly insulted their whole nation. The illiberal reflections cast

---

[1]*placemen:* political appointees to public office

by some Americans on the whole body of the Scotch, as favourers of arbitrary power, restrained high-spirited individuals of that nation from joining a people who suspected their love of liberty. Such of them as adhered to the cause of independence, were steady in their attachment. The army and the Congress ranked among their best officers, and most valuable members, some individuals of that nation.

Such of the Germans in America as possessed the means of information, were generally determined Whigs; but many of them were too little informed to be able to chuse their side on proper ground. They, especially such of them as resided in the interior country, were, from their not understanding the English language, far behind most of the other inhabitants in a knowledge of the merits of the dispute. Their disaffection was rather passive than active: a considerable part of it arose from principles of religion, for some of their sects deny the lawfulness of war. No people have prospered more in America than the Germans. None have surpassed, and but few have equalled them in industry and other republican virtues.

The great body of Tories in the southern States was among the settlers on their western frontier: many of these were disorderly persons, who had fled from the old settlements to avoid the restraints of civil government. Their numbers were increased by a set of men called regulators. The expence and difficulty of obtaining the decision of courts against horse-thieves and other criminals, had induced sundry persons, about the year 1770, to take the execution of the laws into their own hands, in some of the remote settlements, both of North and South-Carolina. In punishing crimes, forms as well as substance must be regarded. From not attending to the former, some of these regulators, though perhaps aiming at nothing but what they thought right, committed many offences both against law and justice. By their violent proceedings regular government was prostrated: they drew on them the vengeance of royal governors: the regulators having suffered from their hands, were slow to oppose an established government, whose power to punish they had recently experienced. Apprehending that the measures of Congress were like their own regulating schemes, and fearing that they would terminate in the same disagreeable consequences, they and their adherents were generally opposed to the revolution.

Religion also divided the inhabitants of America: the Presbyterians and Independents were almost universally attached to the measures of Congress. Their religious societies are governed on the republican plan.

From independence they had much to hope, but from Great Britain, if finally successful, they had reason to fear the establishment of a church hierarchy. Most of the episcopal ministers of the northern provinces were pensioners on the bounty of the British government. The greatest part of their clergy, and many of their laity in these provinces, were therefore disposed to support a connection with Great Britain. The episcopal clergy in the southern provinces being under no such bias, were often among the warmest Whigs. Some of them foreseeing the downfall of religious establishments from the success of the Americans, were less active: but in general, where their church was able to support itself, their clergy and laity zealously espoused the cause of independence. Great pains were taken to persuade them, that those who had been called dissenters, were aiming to abolish the episcopal establishment to make way for their own exaltation; but the good sense of the people restrained them from giving any credit to the unfounded suggestion. Religious controversy was happily kept out of view: the well-informed of all denominations were convinced, that the contest was for their civil rights, and therefore did not suffer any other considerations to interfere, or disturb their union.

The Quakers, with a few exceptions, were averse to independence. In Pennsylvania they were numerous, and had power in their hands. Revolutions in government are rarely patronised by any body of men, who foresee that a diminution of their own importance is likely to result from the change. Quakers from religious principles were averse to war, and therefore could not be friendly to a revolution, which could only be effected by the sword. Several individuals separated from them on account of their principles, and following the impulse of their inclinations, joined their countrymen in arms. The services America received from two of their society, Gen. Greene and Mifflin, made some amends for the embarrassments which the disaffection of the great body of their people occasioned to the exertions of the active friends of independence.

The age and temperament of individuals had often an influence in fixing their political character. Old men were seldom warm Whigs: they could not relish the great changes which were daily taking place; attached to ancient forms and habits, they could not readily accommodate themselves to new systems. Few of the very rich were active in forwarding the revolution. This was remarkably the case in the eastern and middle States; but the reverse took place in the southern extreme of the confederacy. There were in no part of America more determined Whigs than the opulent slaveholders in Virginia, the Carolinas, and Georgia. The active and spirited part of the community, who felt themselves possessed of talents that would raise them to eminence in a free government, longed for the establishment of independent constitutions: but those who were in possession or expectation of royal favour, or of promotion from Great Britain, wished that the connection between the Parent State and the Colonies might be preserved. The young, the ardent, the ambitious, and the enterprising, were mostly Whigs; but the phlegmatic, the timid, the interested, and those who wanted decision were, in general, favourers of Great Britain, or at least only the lukewarm, inactive friends of independence. The Whigs received a great re-inforcement from the operation of continental money. In the year 1775, 1776, and in the first months of 1777, while the bills of Congress were in good credit, the effects of them were the same, as if a foreign power had made the United States a present of twenty million of silver dollars. The circulation of so large a sum of money, and the employment given to great numbers in providing for the American army, increased the numbers and invigorated the zeal of the friends of the revolution; on the same principles, the American war was patronised in England, by the many contractors and agents for transporting and supplying the British army. In both cases the inconveniencies of interrupted commerce were lessened by the employment which war and a domestic circulation of money submitted in its room. The convulsions of war afforded excellent shelter for desperate debtors. The spirit of the times revolted against dragging to jails for debt, men who were active and zealous in defending their country, and on the other hand, those who owed more than they were worth, by going within the British lines, and giving themselves the merit of suffering on the score of loyalty, not only put their creditors to defiance, but sometimes obtained promotion, or other special marks of royal favour.

The American revolution, on the one hand, brought forth great vices; but on the other hand, it called forth many virtues, and gave occasion for the display of abilities which, but for that event, would have been lost to the world. When the war began, the Americans were a mass of husbandmen, merchants, mechanics, and fishermen; but the necessities of the country gave a spring to the active powers of the inhabitants, and set them on thinking, speaking, and acting, in a line far beyond that to which they had been accustomed. The difference between nations is not so much owing to nature, as to education and circumstances. While the Americans were guided by the leading strings of the Mother Country, they had no scope nor encouragement for exertion. All the departments of government were established and executed for them, but not by them. In the years 1775 and 1776, the country, being suddenly thrown into a situation that needed the abilities of all its sons, these generally took their places, each according to the bent of his inclination. As they severally pursued their objects with ardour, a vast expansion of the human mind speedily followed. This displayed itself in a variety of ways. It was found that the talents for great stations did not differ in kind, but only in degree, from those which were necessary for the proper discharge of the ordinary business of civil society. In the bustle that was occasioned by the war, few instances could be produced of any persons who made a figure, or who rendered essential services, but from among those who had given specimens of similar talents in their respective professions. Those who, from indolence or dissipation, had been of little service to the community in time of peace, were found equally unserviceable in war. A few young men were exceptions to this general rule. Some of these, who had indulged in youthful follies, broke off from their vicious courses, and on the pressing call of their country became useful servants of the public: but the great bulk of those, who were the active instruments of carrying on the revolution, were self-made, industrious men. These who by their own exertions had established or laid a foundation for establishing personal independence, were most generally trusted, and most successfully employed in establishing that of their country. In these times of action, classical education was found of less service than good natural parts, guided by common sense and sound judgement.

Several names could be mentioned of individuals who, without the knowledge of any other language than their mother tongue, wrote not only accurately, but elegantly on public business. It seemed as if the war not only required, but created talents. Men whose minds were warmed with the love of liberty, and whose abilities were improved by daily exercise, and sharpened with a laudable ambition to serve their distressed country, spoke, wrote, and acted, with an energy far surpassing all expectations which could be reasonably founded on their previous acquirements.

The Americans knew but little of one another, previous to the revolution. Trade and business had brought the inhabitants of their seaports acquainted with each other, but the bulk of the people in the interior country were unacquainted with their fellow-citizens. A continental army, and a Congress composed of men from all the States, by freely mixing together, were assimilated into the mass. Individuals of both, mingling with the citizens, disseminated principles of union among them. Local prejudices abated. By frequent collision asperities were worn off, and a foundation was laid for the establishment of a nation out of discordant materials. Intermarriages between men and women of different States were much more common than before the war, and became an additional cement to the union. Unreasonable jealousies had existed between the inhabitants of the eastern and of the southern States; but on becoming better acquainted with each other, these in a great measure subsided. A wiser policy prevailed. Men of liberal minds led the way in discouraging local distinctions, and the great body of the people, as soon as reason got the better of prejudice, found that their best interests would be most effectually promoted by such practices and sentiments as were favourable to union. . . .

Such have been some of the beneficial effects which have resulted from that expansion of the human mind, which has been produced by the revolution; but these have not been without alloy.

To overset an established government, unhinges many of those principles which bind individuals to each other. A long time, and much prudence, will be necessary to re-produce a spirit of union and that reverence for government, without which society is a rope of sand. The right of people to resist their rulers, when invading their liberties, forms the corner stone of the American republics. This principle, though just in itself, is not favourable to the tranquillity of present establishments. The maxims and measures, which in the years 1774 and 1775 were successfully inculcated and adopted by American patriots, for oversetting the established government, will answer a similar purpose, when recurrence is had to them by factious demagogues for disturbing the freest government that were ever devised.

War never fails to injure the morals of the people engaged in it. The American war, in particular, had an unhappy influence of this kind. Being begun without funds or regular establishments, it could not be carried on without violating private rights; and in its progress, it involved a necessity for breaking solemn promises, and plighted public faith. The failure of national justice, which was in some degree unavoidable, increased the difficulties of performing private engagements, and weakened that sensibility to the obligations of public and private honour, which is a security for the punctual performance of contracts. . . .

On the whole, the literary, political, and military talents of the citizens of the United States have been improved by the revolution, but their moral character is inferior to what it formerly was. So great is the change for the worse, that the friends of public order were loudly called upon to exert their utmost abilities in extirpating the vicious principles and habits which have taken deep root during the late convulsions.

CHAP. XXVII

The Discharge of the American Army; the Evacuation of New-York: the Resignation of General Washington; Arrangements of Congress for the disposing of their western Territory, and paying their Debts; the Distresses of the States after the Peace; the inefficacy of the Articles of the Confederation; a grand Convention for amending the Government; the new Constitution; General Washington appointed President; an Address to the People of the United States.

. . . When the people on the return of peace supposed their troubles to be ended, they found them to be only varied. The calamities of war were followed by another class of evils, different in their origin, but not less injurious in their consequences. The inhabitants feeling the pressure of their sufferings, and not knowing precisely from what source they originated, or how to remedy them, became uneasy, and many were ready to adopt any desperate measures that turbulent leaders might recommend. In this irritable state, a great number of the citizens of Massachusetts, sore with their enlarged portion of public calamity, were induced by seditious demagogues to make an open resistance to the operations of their own free government. Insurrections took place in many parts, and laws were trampled upon by the very men whose deputies had enacted them, and whose deputies might have repealed them. By the moderation of the legislature, and especially by the bravery and good conduct of Generals Lincoln and Shepherd, and the firmness of the well-affected militia, the insurgents were speedily quelled, and good order restored, with the loss of about six of the freemen of the State.

The untoward events which followed the re-establishment of peace, though evils of themselves, were overruled for great national good. From the failure of their expectations of an immediate increase of political happiness, the lovers of liberty and independence began to be less sanguine in their hopes from the American revolution, and to fear that they had built a visionary fabric of government on the fallacious ideas of public virtue; but that elasticity of the human mind, which is nurtured by free constitutions, kept them from desponding. By an exertion of those inherent principles of self-preservation which republics possess, a recurrence was had to the good sense of the people for the rectification of fundamental disorders. While the country, free from foreign force and domestic violence, enjoyed tranquillity, a proposition was made by Virginia to all the other States to meet in convention, for the purpose of digesting a form of government, equal to the exigencies of the union. The first motion for this purpose was made by Mr. Madison, and he had the pleasure of seeing it acceded to by twelve of the States, and finally to issue in the establishment of a New Constitution, which bids fair to repay the citizens of the United States for the toils, dangers, and wastes of the revolution. . . .

The new constitution having been ratified by eleven of the States, and senators and representatives having been chosen agreeably to the articles thereof, they met at New-York and commenced proceedings under it. The old Congress and confederation, like the continental money, expired without a sigh or groan. A new Congress, with more ample powers, and a new constitution, partly national, and partly fœderal, succeeded in their place, to the great joy of all who wished for the happiness of the United States.

Though great diversity of opinions have prevailed about the new constitution, there was but one opinion about the person who should be appointed its supreme executive officer. The people, as well antifœderalists as fœderalists, (for by these names the parties for and against the new constitution were called) unanimously turned their eyes on the late commander of their armies, as the most proper person to be their first President. Perhaps there was not a well-informed individual in the United States, (Mr. Washington himself only excepted) who was not anxious that he should be called to the executive administration of the proposed new plan of government. Unambitious of farther honours he had retired to his farm in Virginia, and hoped to be excused from all farther public service; but his country called him by an unanimous vote to fill the highest station in its gift. That honest zeal for the public good, which had uniformly influenced him to devote both his time and talents to the service of his country, got the better of his love of retirement, and induced him once more to engage in the great business of making a nation happy. The intelligence of his election being communicated to him, while on his farm in

Virginia, he set out soon after for New-York. On his way thither, the road was crowded with numbers anxious to see the Man of the people. Escorts of militia, and of gentlemen of the first character and station, attended him from State to State, and he was every where received with the highest honours which a grateful and admiring people could confer. Addresses of congratulation were presented to him by the inhabitants of almost every place of consequence through which he passed, to all of which he returned such modest, unassuming answers as were in every respect suitable to his situation. So great were the honours with which he was loaded, that they could scarcely have failed to produce haughtiness in the mind of any ordinary man; but nothing of the kind was ever discovered in this extraordinary personage. On all occasions he behaved to all men with the affability of one citizen to another. He was truly great in deserving the plaudits of his country, but much greater in not being elated with them. . . .

Citizens of the United States! you have a well-balanced constitution established by general consent, which is an improvement on all republican forms of government heretofore established. It possesses the good qualities of monarchy, but without its vices. The wisdom and stability of an aristocracy, but without the insolence of hereditary masters. The freedom and independence of a popular assembly, acquainted with the wants and wishes of the people, but without the capacity of doing those mischiefs which result from uncontrolled power in one assembly. The end and object of it is public good. If you are not happy it will be your own fault. No knave or fool can plead an hereditary right to sport with your property or your liberties. Your laws and your law-givers must all proceed from yourselves. You have the experience of nearly six thousand years, to point out the rocks on which former republics have been dashed to pieces. Learn wisdom from their misfortunes. Cultivate justice both public and private. No government will or can endure which does not protect the rights of its subjects. Unless such efficient regulations are adopted, as will secure property as well as liberty, one revolution will follow another. Anarchy, monarchy, or despotism, will be the consequence. By just laws and the faithful execution of them, public and private credit will be restored, and the restoration of credit will be a mine of wealth to this young country. It will make a fund for agriculture, commerce and manufactures, which will soon enable the United States to claim an exalted rank among the nations of the earth. Such are the resources of your country, and so trifling are your debts, compared with your resources, that proper systems, wisely planned and faithfully executed, will soon fill your extensive territory with inhabitants, and give you the command of such ample capitals, as will enable you to run the career of national greatness, with advantages equal to the oldest kingdoms of Europe. What they have been slowly growing to, in the course of near two thousand years, you may hope to equal within one century. If you continue under one government, built on the solid foundations of public justice, and public virtue, there is no point of national greatness to which you may not aspire with a well-founded hope of speedily attaining it. Cherish and support a reverence for government, and cultivate an union between the East and South, the Atlantic and the Mississippi. Let the greatest good of the greatest number be the pole-star of your public and private deliberations. Shun wars, they beget debt, add to the common vices of mankind, and produce others, which are almost peculiar to themselves. Agriculture, manufactures, and commerce, are your proper business. Seek not to enlarge your territory by conquest; it is already sufficiently extensive. You have ample scope for the employment of your most active minds, in promoting your own domestic happiness. Maintain your own rights, and let all others remain in quiet possession of theirs. Avoid discord, faction, luxury, and the other vices which have been the bane of commonwealths. Cherish and reward the philosophers, the statesmen, and the patriots, who devote their talents and time, at the expence of their private interests, to the toils of enlightening and directing their fellow citizens, and thereby rescue citizens and rulers of republics from the common and too often merited charge of ingratitude. Practise industry, frugality, temperance, moderation, and the whole lovely train of republican virtues. Banish from your borders the liquid fire of the West-Indies, which, while it entails poverty and disease, prevents industry, and foments private quarrels. Venerate the plough, the hoe, and all the implements of agriculture. Honour the men, who with their own hands maintain their families, and raise up children who are inured to toil, and capable of defending their country. Reckon the necessity of labour not among the curses, but the blessings of life. Your towns will probably ere long be engulphed in luxury and effeminacy. If your liberties and future prospects depend on them, your career of liberty would probably be short; but a great majority of your country, must, and will be yeomanry, who have no other dependence than on Almighty God for his usual blessing on their daily labour. From the great excess of the numbers of such independent farmers in these States, over and above all other classes of inhabitants, the long continuance of your liberties may be reasonably presumed.

Let the hapless African sleep undisturbed on his native shore, and give over wishing for the extermination of the ancient proprietors of this land. Universal justice is universal interest. The most enlarged happiness of one people, by no means requires the degradation or destruction of another. It would be more glorious to civilise one tribe of savages than to exterminate or expel a score. There is territory enough for them and for you. Instead of invading their rights, promote their happiness, and give them no reason to curse the folly of their fathers, who suffered your's to sit down on a soil which the common Parent of us both had previously assigned to them: but above all, be particularly careful that your own descendants do not degenerate into savages. Diffuse the means of education, and particularly of religious instruction, through your remotest settlements. To this end, support and strengthen the hands of public teachers, and especially of worthy clergymen. Let your voluntary contributions confute the dishonourable position, that religion cannot be supported but by compulsory establishments. Remember that there can be no political happiness without liberty; that there can be no liberty without morality; and that there can be no morality without religion.

It is now your turn to figure on the face of the earth, and in the annals of the world. You possess a country which in less than a century will probably contain fifty millions of inhabitants. You have, with a great expence of blood and treasure, rescued yourselves and your posterity from the domination of Europe. Perfect the good work you have begun, by forming such arrangements and institutions as bid fair for ensuring to the present and future generations the blessings for which you have successfully contended.

May the Almighty Ruler of the Universe, who has raised you to independence, and given you a place among the nations of the earth, make the American Revolution an era in the history of the world, remarkable for the progressive increase of human happiness!

14

MERCY OTIS WARREN

History of the Rise, Progress and Termination of the American Revolution

1805

One of the leading female intellectuals and most prolific authors of the early Republic, Mercy Otis Warren (1728–1814) was a member of New England's social and political elite. Her History of the Rise, Progress and Termination of the American Revolution, *begun during the Revolution, was all but finished by 1791 and finally published in Boston in 1805. Reviews of the book were highly partisan: whereas a Jeffersonian paper in Massachusetts praised Warren's character sketches and style, Federalist publications suggested that she heed the first Epistle of Paul and "keep silent." Her old friend John Adams was particularly vituperative about her portrait of his "lapse from former republican principles." He concluded that "History is not the Province of the Ladies."*

CHAPTER IV

... It is ever painful to a candid mind to exhibit the deformed features of its own species; yet truth requires a just portrait of the public delinquent, though he may possess such a share of private virtue as would lead us to esteem the man in his domestic character, while we detest his political, and execrate his public transactions.

The barriers of the British constitution broken over, and the ministry encouraged by their sovereign, to pursue the iniquitous system against the colonies to the most alarming extremities, they probably judged it a prudent expedient, in order to curb the refractory spirit of the Massachusetts, perhaps bolder in sentiment and earlier in opposition than some of the other colonies, to appoint a man to preside over them who had renounced the *quondam* ideas of public virtue, and sacrificed all principle of that nature on the altar of ambition.

Mercy Otis Warren, *History of the Rise, Progress and Termination of the American Revolution* (Boston, 1805), 3 vols., 1:78–79, 81–96, 101, 105–11.

Soon after the recal[1] of Mr. Bernard, Thomas Hutchinson, Esq. a native of Boston, was appointed to the government of Massachusetts. All who yet remember his pernicious administration and the fatal consequences that ensued, agree, that few ages have produced a more fit instrument for the purposes of a corrupt court. He was dark, intriguing, insinuating, haughty and ambitious, while the extreme of avarice marked each feature of his character. His abilities were little elevated above the line of mediocrity; yet by dint of industry, exact temperance, and indefatigable labor, he became master of the accomplishments necessary to acquire popular fame. Though bred a merchant, he had looked into the origin and the principles of the British constitution, and made himself acquainted with the several forms of government established in the colonies; he had acquired some knowledge of the *common law* of England, diligently studied the intricacies of *Machiavel[l]ian* policy, and never failed to recommend the Italian master as a model to his adherents. . . .

Mr. Hutchinson was one of the first in America who felt the full weight of popular resentment. His furniture was destroyed, and his house levelled to the ground, in the tumults occasioned by the news of the stamp-act. Ample compensation was indeed afterwards made him for the loss of property, but the strong prejudices against his political character were never eradicated.

All pretences to moderation on the part of the British government now laid aside, the full appointment of Mr. Hutchinson to the government of the Massachusetts was publickly announced at the close of the year one thousand seven hundred and sixty-nine. On his promotion the new governor uniformly observed a more high-handed and haughty tone than his predecessor. He immediately, by an explicit declaration, avowed his independence on the people, and informed the legislative that his majesty had made ample provision for his support without their aid or suffrages. The vigilant guardians of the rights of the people directly called upon him to relinquish the unconstitutional stipend, and to accept the free grants of the general assembly for his subsistence, as usually practised. He replied that an acceptance of this offer would be a breach of his instructions from the king. This was his constant apology for every arbitrary step.

Secure of the favor of his sovereign, and now regardless of the popularity he had formerly courted with such avidity, he decidedly rejected the idea of responsibility to, or dependence on, the people. With equal inflexibility he disregarded all arguments used for the removal of the troops from the capital, and permission to the council and house of representatives to return to the usual seat of government. He silently heard their solicitations for this purpose, and as if with a design to pour contempt on their supplications and complaints, he within a few days after withdrew a garrison, in the pay of the province, from a strong fortress in the harbour of Boston; placed two regiments of the king's troops in their stead, and delivered the keys of the castle to colonel Dalrymple, who then commanded the king's troops through the province.

These steps, which seemed to bid defiance to complaint, created new fears in the minds of the people. It required the utmost vigilance to quiet the murmurs and prevent the fatal consequences apprehended from the ebullitions of popular resentment. But cool, deliberate and persevering, the two houses continued to resolve, remonstrate, and protest, against the infractions on their charter, and every dangerous innovation on their rights and privileges. Indeed the intrepid and spirited conduct of those, who stood forth undaunted at this early crisis of hazard, will dignify their names so long as the public records shall remain to witness their patriotic firmness.

Many circumstances rendered it evident that the ministerial party wished a spirit of opposition to the designs of the court might break out into violence, even at the expense of blood. This they thought would in some degree have sanctioned a measure suggested by one of the faction in America, devoted to the arbitrary system, "That some method must be devised, to take off the original *incendiaries** whose writings instilled the poison of sedition through the vehicle of the Boston Gazette"†

Had this advice been followed, and a few gentlemen of integrity and ability, who had spirit sufficient to make an effort in favor of their country in each colony, have been seized at the same moment and immolated early in the contest on the bloody altar of power, perhaps Great Britain might have held the continent in subjection a few years longer.

*See Andrew Oliver's letter to one of the ministry, dated February 13, 1769.

†This gazette was much celebrated for the freedom of its disquisitions in favor of civil liberty. It has been observed that

"it will be a treasury of political intelligence for the historians of this country. [James] Otis, [Oxenbridge] Thacher, [Samuel] Dexter, [John] Adams, [James] Warren, and [Josiah] Quincy, Doctors Samuel Cooper and [Jonathan] Mayhew, stars of the first magnitude in our northern hemisphere, whose glory and brightness distant ages will admire; these gentlemen of character and influence offered their first essays to the public through the medium of the Boston Gazette, on which account the paper became odious to the friends of prerogative, but not more disgusting to the tories and high church than it was pleasing to the whigs."

That they had measures of this nature in contemplation there is not a doubt. Several instances of a less atrocious nature confirmed this opinion, and the turpitude of design which at this period actuated the court party was clearly evinced by the attempted assassination of the celebrated Mr. Otis, justly deemed the first martyr to American freedom; and truth will enrol[l] his name among the most distinguished patriots who have expired on the "blood-stained theatre of human action."

This gentleman, whose birth and education was equal to any in the province, possessed an easy fortune, independent principles, a comprehensive genius, strong mind, retentive memory, and great penetration. To these endowments may be added that extensive professional knowledge, which at once forms the character of the complete civilian and the able statesman.

In his public speeches, the fire of eloquence, the acumen of argument, and the lively sallies of wit, at once warmed the bosom of the stoic and commanded the admiration of his enemies. To his probity and generosity in the public walks were added the charms of affability and improving converse in private life. His humanity was conspicuous, his sincerity acknowledged, his integrity unimpeached, his honor unblemished, and his patriotism marked with the disinterestedness of the Spartan. Yet he was susceptible of quick feelings and warm passions, which in the ebullitions of zeal for the interest of his country sometimes betrayed him into unguarded epithets that gave his foes an advantage, without benefit to the cause that lay nearest his heart.

He had been affronted by the partizans of the crown, vilified in the public papers, and treated (after his resignation of office*) in a manner too gross for a man of his spirit to pass over with impunity. Fearless of consequences, he had always given the world his opinions both in his writings and his conversation, and had recently published some severe strictures on the conduct of the commissioners of the customs and others of the ministerial party, and bidding defiance to resentment, he supported his allegations by the signature of his name.

A few days after this publication appeared, Mr. Otis with only one gentleman in company was suddenly assaulted in a public room, by a band of ruffians armed with swords and bludgeons. They were headed by John Robinson, one of the commissioners of the customs. The lights were immediately extinguished, and Mr. Otis covered with

*Office of judge advocate in governor Bernard's administration

wounds was left for dead, while the assassins made their way through the crowd which began to assemble; and before their crime was discovered, fortunately for themselves, they escaped soon enough to take refuge on board one of the king's ships which then lay in the harbor.

In a state of nature, the savage may throw his poisoned arrow at the man, whose soul exhibits a transcript of benevolence that upbraids his own ferocity, and may boast his blood-thirsty deed among the hordes of the forest without disgrace; but in a high stage of civilization, where humanity is cherished, and politeness is become a science, for the dark assassin then to level his blow at superior merit, and screen himself in the arms of power, reflects an odium on the government that permits it, and puts human nature to the blush.

The party had a complete triumph in this guilty deed; for though the wounds did not prove mortal, the consequences were tenfold worse than death. The future usefulness of this distinguished *friend* of his country was destroyed, reason was shaken from its throne, genius obscured, and the great man in ruins lived several years for his friends to weep over, and his country to lament the deprivation of talents admirably adapted to promote the highest interests of society.

This catastrophe shocked the feelings of the virtuous not less than it raised the indignation of the brave. Yet a remarkable spirit of forbearance continued for a time, owing to the respect still paid to the opinions of this unfortunate gentleman, whose voice though always opposed to the strides of despotism was ever loud against all tumultuous and illegal proceedings. He was after a partial recovery sensible himself of his incapacity for the exercise of talents that had shone with peculiar lustre, and often invoked the messenger of death to give him a sudden release from a life become burdensome in every view but when the calm interval of a moment permitted him the recollection of his own integrity. In one of those intervals of beclouded reason he forgave the murderous band, after the principal ruffian had asked pardon in a court of justice;* and at the intercession of the gentleman whom he had so grossly abused, the people forebore inflicting that summary vengeance which was generally thought due to so black a crime.

Mr. Otis lived to see the independence of America, though in a state of mind incapable of enjoying fully the glorious event which his

---

*On a civil process commenced against him, John Robinson was adjudged to pay five thousand pounds sterling damages; but Mr. Otis despising all pecuniary compensation, relinquished it on the culprit's asking pardon and setting his signature on a very humble acknowledgment.

own exertions had precipitated. After several years of mental derange-
ment, as if in consequence of his own prayers, his great soul was
instantly set free by a flash of lightning, from the evils in which the
love of his country had involved him. His death took place in May, one
thousand seven hundred and eighty three, the same year the peace
was concluded between Great Britain and America.*

Though the parliamentary system of colonial regulations was in
many instances similar, and equally aimed to curtail the privileges of
each province, yet no military force had been expressly called in aid of
civil authority in any of them, except the Massachusetts. From this circumstance some began to flatter themselves that more lenient dispositions were operating in the mind of the king of Great Britain, as well
as in the parliament and the people towards America in general. . . .

In the mean time the inhabitants of the town of Boston had suffered almost every species of insult from the British soldiery; who,
countenanced by the royal party, had generally found means to screen
themselves from the hand of the civil officers. Thus all authority
rested on the point of the sword, and the partizans of the crown tri-
umphed for a time in the plenitude of military power. Yet the measure
and the manner of posting troops in the capital of the province, had
roused such jealousy and disgust, as could not be subdued by the
scourge that hung over their heads. Continual bickerings took place in
the streets between the soldiers and the citizens; the insolence of the
first, which had been carried so far as to excite the African slaves to

*A sister touched by the tenderest feelings, while she has thought it her duty to do
justice to a character neglected by some, and misrepresented by other historians, can
exculpate herself from all suspicion of partiality by the testimony of many of his countrymen who witnessed his private merit and public exertions. But she will however only
subjoin a paragraph of a letter written to the author of these annals, on the news of Mr.
Otis's death, by John Adams, Esq. then minister plenipotentiary from the United States
to the court of France.

PARIS, SEPTEMBER 10TH, 1783.
"It was, Madam, with very afflicting sentiments I learned the death of Mr. Otis, my
worthy master. Extraordinary in death as in life, he has left a character that will never
die while the memory of the American revolution remains; whose foundation he laid
with an energy, and with those masterly abilities, which no other man possessed."

The reader also may not be displeased at an extemporary exclamation of a gentleman of
poetic talents, on hearing of the death of Mr. Otis.

When God in anger saw the spot,
On earth to Otis given,
In thunder as from Sinai's mount,
He snatch'd him back to heaven.

murder their masters, with the promise of impunity,* and the indiscretion of the last, was often productive of tumults and disorder that led the most cool and temperate to be apprehensive of consequences of the most serious nature.

No previous outrage had given such a general alarm, as the commotion on the fifth of March, one thousand seven hundred and seventy. Yet the accident that created a resentment which emboldened the timid, determined the wavering, and awakened an energy and decision that neither the artifices of the courtier, nor the terror of the sword could easily overcome, arose from a trivial circumstance; a circumstance which but from the consideration that these minute accidents frequently lead to the most important events, would be beneath the dignity of history to record.

A centinel posted at the door of the custom house had seized and abused a boy, for casting some opprobrious reflections on an officer of rank; his cries collected a number of other lads, who took the childish revenge of pelting the soldier with snow-balls. The main-guard stationed in the neighborhood of the custom-house, was informed by some persons from thence, of the rising tumult. They immediately turned out under the command of a captain Preston, and beat to arms. Several *fracas* of little moment had taken place between the soldiery and some of the lower class of inhabitants, and probably both were in a temper to avenge their own private wrongs. The cry of fire was raised in all parts of the town, the mob collected, and the soldiery from all quarters ran through the streets sword in hand, threatening and wounding the people, and with every appearance of hostility, they rushed furiously to the centre of the town.

The soldiers thus ready for execution, and the populace grown outrageous, the whole town was justly terrified by the unusual alarm. This naturally drew out persons of higher condition, and more peaceably disposed, to inquire the cause. Their consternation can scarcely be described, when they found orders were given to fire promiscuously among the unarmed multitude. Five or six persons fell at the first fire, and several more were dangerously wounded at their own doors.

These sudden popular commotions are seldom to be justified, and their consequences are ever to be dreaded. It is needless to make any

---

*Capt. Wilson of the 29th regiment was detected in the infamous practice; and it was proved beyond a doubt by the testimony of some respectable citizens, who declared on oath, that they had accidentally witnessed the offer of reward to the blacks, by some subaltern officers, if they would rob and murder their masters.

observations on the assumed rights of royalty, in a time of peace to disperse by military murder the disorderly and riotous assemblage of a thoughtless multitude. The question has frequently been canvassed; and was on this occasion thoroughly discussed, by gentlemen of the first professional abilities.

The remains of loyalty to the sovereign of Britain were not yet extinguished in American bosoms, neither were the feelings of compassion, which shrunk at the idea of human carnage, obliterated. Yet this outrage enkindled a general resentment that could not be disguised; but every method that prudence could dictate, was used by a number of influential gentlemen to cool the sudden ferment, to prevent the populace from attempting immediate vengeance, and to prevail on the multitude to retire quietly to their own houses, and wait the decisions of law and equity. They effected their humane purposes; the people dispersed; and captain Preston and his party were taken into custody of the civil magistrate. A judicial inquiry was afterwards made into their conduct; and so far from being actuated by any partial or undue bias, some of the first counsellors at law engaged in their defence; and after a fair and legal trial they were acquitted of premeditated or wilful murder, by a jury of the county of Suffolk.

The people, not dismayed by the blood of their neighbors thus wantonly shed, determined no longer to submit to the insolence of military power. Colonel Dalrymple, who commanded in Boston, was informed the day after the riot in King Street, "that he must withdraw his troops from the town within a limited term, or hazard the consequences."

The inhabitants of the town assembled in Faneuil Hall, where the subject was discussed with becoming spirit, and the people unanimously resolved, that no armed force should be suffered longer to reside in the capital; that if the king's troops were not immediately withdrawn by their own officers, the governor should be requested to give orders for their removal, and thereby prevent the necessity of more rigorous steps. A committee from the body was deputed to wait on the governor, and request him to exert that authority which the exigencies of the times required from the supreme magistrate. Mr. Samuel Adams, the chairman of the committee, with a pathos and address peculiar to himself, exposed the illegality of quartering troops in the town in the midst of peace; he urged the apprehensions of the people, and the fatal consequences that might ensue if their removal was delayed.

But no arguments could prevail on Mr. Hutchinson; who either from timidity, or some more censurable cause, evaded acting at all in the business, and grounded his refusal on a pretended want of authority. After which, colonel Dalrymple, wishing to compromise the matter, consented that the twenty-ninth regiment, more culpable than any other in the late tumult, should be sent to Castle Island. This concession was by no means satisfactory; the people, inflexible in their demands, insisted that not one British soldier should be left within the town; their requisition was reluctantly complied with, and within four days the whole army decamped. It is not to be supposed, that this compliance of British veterans originated in their fears of an injured and incensed people, who were not yet prepared to resist by arms. They were undoubtedly sensible they had exceeded their orders, and anticipated the designs of their master; they had rashly begun the slaughter of Americans, and enkindled the flames of civil war in a country, where allegiance had not yet been renounced. . . .

It has already been observed, that the revenue acts which had occasioned a general murmur, had been repealed, except a small duty on all India teas, by which a claim was kept up to tax the colonies at pleasure, whenever it should be thought expedient. This was an article used by all ranks in America; a luxury of such universal consumption, that administration was led to believe, that a monopoly of the sales of tea, might be so managed, as to become a productive source of revenue. . . .

The storage or detention of a few cargoes of teas is not an object in itself sufficient to justify a detail of several pages; but as the subsequent severities towards the Massachusetts were grounded on what the ministry termed their *refractory behavior* on this occasion; and as those measures were followed by consequences of the highest magnitude both to Great Britain and the colonies, a particular narration of the transactions of the town of Boston is indispensable. There the sword of civil discord was first drawn, which was not re-sheathed until the emancipation of the thirteen colonies from the yoke of foreign domination was acknowledged by the diplomatic seals of the first powers in Europe. This may apologize, if necessary, for the appearance of locality in the preceding pages, and for its farther continuance in regard to a colony, on which the bitterest cup of ministerial wrath was poured for a time, and where the energies of the human mind were earlier called forth, than in several of the sister states.

Not intimidated by the frowns of greatness, nor allured by the smiles of intrigue, the vigilance of the people was equal to the importance of the event. Though expectation was equally awake in both parties, yet three or four weeks elapsed in a kind of *inertia;* the one side flattered themselves with hopes, that as the ships were suffered to be so long unmolested, with their cargoes entire, the point might yet be obtained; the other thought it possible, that some impression might yet be made on the governor, by the strong voice of the people.

Amidst this suspense a rumour was circulated, that admiral Montague was about to seize the ships, and dispose of their cargoes at public auction, within twenty-four hours. This step would as effectually have secured the duties, as if sold at the shops of the consignees, and was judged to be only a *finesse*, to place them there on their own terms. On this report, convinced of the necessity of preventing so bold an attempt, a vast body of people convened suddenly and repaired to one of the largest and most commodious churches in Boston; where, previous to any other steps, many fruitless messages were sent both to the governor and the consignees, whose timidity had prompted them to a seclusion from the public eye. Yet they continued to refuse any satisfactory answer; and while the assembled multitude were in quiet consultation on the safest mode to prevent the sale and consumption of an herb, *noxious* at least to the political constitution, the debates were interrupted by the entrance of the sheriff with an order from the governor, styling them an illegal assembly, and directing their immediate dispersion.

This authoritative mandate was treated with great contempt, and the sheriff instantly hissed out of the house. A confused murmur ensued, both within and without the walls; but in a few moments all was again quiet, and the leaders of the people returned calmly to the point in question. Yet every expedient seemed fraught with insurmountable difficulties, and evening approaching without any decided resolutions, the meeting was adjourned without day.

Within an hour after this was known abroad, there appeared a great number of persons, clad like the aborigines of the wilderness, with tomahawks in their hands, and clubs on their shoulders, who without the least molestation marched through the streets with silent solemnity, and amidst innumerable spectators, proceeded to the wharves, boarded the ships, demanded the keys, and with much deliberation knocked open the chests, and emptied several thousand weight of the finest teas into the ocean. No opposition was made, though surrounded by the king's ships; all was silence and dismay.

This done, the procession returned through the town in the same order and solemnity as observed in the outset of their attempt. No other disorder took place, and it was observed, the stillest night ensued that Boston had enjoyed for many months. This unexpected event struck the ministerial party with rage and astonishment; while, as it seemed to be an attack upon private property, many who wished well to the public cause could not fully approve of the measure. Yet perhaps the laws of self-preservation might justify the deed, as the exigencies of the times required extraordinary exertions, and every other method had been tried in vain, to avoid this disagreeable alternative. Besides it was alleged, and doubtless it was true, the people were ready to make ample compensation for all damages sustained, whenever the unconstitutional duty should be taken off, and other grievances radically redressed. But there appeared little prospect that any conciliatory advances would soon be made. The officers of government discovered themselves more vindictive than ever: animosities daily increased, and the spirits of the people were irritated to a degree of alienation, even from their tenderest connexions, when they happened to differ in political opinion.

By the frequent dissolution of the general assemblies, all public debate had been precluded, and the usual regular intercourse between the colonies cut off. The modes of legislative communication thus obstructed, at a period when the necessity of harmony and concert was obvious to every eye, no systematical opposition to gubernatorial intrigues, supported by the king and parliament of Great Britain, was to be expected without the utmost concord, confidence, and union of all the colonies. Perhaps no single step contributed so much to cement the union of the colonies, and the final acquisition of independence, as the establishment of committees of correspondence. This supported a chain of communication from New Hampshire to Georgia, that produced unanimity and energy throughout the continent.

As in these annals there has yet been no particular mention made of this institution, it is but justice to name at once the author, the origin, and the importance of the measure.

At an early period of the contest, when the public mind was agitated by unexpected events, and remarkably pervaded with perplexity and anxiety, James Warren, Esq. of Plymouth first proposed this institution to a private friend, on a visit at his own house.* Mr. Warren had been an active and influential member of the general assembly from the beginning of the troubles in America, which commenced soon after the demise of George the second. The principles and firmness of

*Samuel Adams, Esq. of Boston

this gentleman were well known, and the uprightness of his character had sufficient weight to recommend the measure. As soon as the proposal was communicated to a number of gentlemen in Boston, it was adopted with zeal, and spread with the rapidity of enthusiasm, from town to town, and from province to province.* Thus an intercourse was established, by which a similarity of opinion, a connexion of interest, and a union of action appeared, that set opposition at defiance, and defeated the machinations of their enemies through all the colonies.

The plan suggested was clear and methodical; it proposed that a public meeting should be called in every town; that a number of persons should be selected by a plurality of voices; that they should be men of respectable characters, whose attachment to the great cause of America had been uniform; that they should be vested by a majority of suffrages with power to take cognizance of the state of commerce, of the intrigues of *toryism,* of litigious ruptures that might create disturbances, and every thing else that might be thought to militate with the rights of the people, and to promote every thing that tended to general utility.

The business was not tardily executed. Committees were every where chosen, who were directed to keep up a regular correspondence with each other, and to give information of all intelligence received, relative to the proceedings of administration, so far as they affected the interest of the British colonies throughout America. The truth was faithfully and diligently discharged, and when afterwards all legislative authority was suspended, the courts of justice shut up, and the last traits of British government annihilated in the colonies, this new institution became a kind of juridical tribunal. Its injunctions were influential beyond the hopes of its most sanguine friends, and the recommendations of committees of correspondence had the force of law. Thus, as despotism frequently springs from anarchy, a regular democracy sometimes arises from the severe encroachments of despotism. . . .

*The general impulse at this time seemed to operate by sympathy, before consultation could be had; thus it appeared afterwards that the vigilant inhabitants of Virginia had concerted a similar plan about the same period.

15

MASON LOCKE WEEMS

The Life of George Washington; with Curious Anecdotes, Equally Honourable to Himself and Exemplary to his Young Countrymen

1809

An Anglican minister in rural Maryland and Virginia, Mason Locke Weems (1759–1825) also worked as an itinerant bookseller for a Philadelphia publisher after 1794. His travels between New York and Georgia taught Weems a great deal about popular tastes and helped him to write one of the best-selling books of the period. His Life of George Washington, *published in Philadelphia in 1800, grew from eighty to two hundred pages by the ninth edition in 1809. Although reviewers scorned it as fiction, the book went through twenty-nine editions by 1825. The fifth edition in 1806 introduced the tale of the cherry tree, creating one of the most enduring and unshakable myths of American culture.*

CHAPTER I

> OH! *as along the stream of time thy name*
> *Expanded flies, and gathers all its fame;*
> *May then these lines to future days descend,*
> *And prove thy* COUNTRY'S *good thine* only *end!*

"Ah, *gentlemen!*"—exclaimed Bonaparte—'twas just as he was about to embark for Egypt . . . some young Americans happening at Toulon, and anxious to see the mighty Corsican, had obtained the honour of an introduction to him. Scarcely were past the customary salutations, when he eagerly asked, *"how fares your countryman, the great* WASHINGTON?" "He was very well," replied the youths, brightening at the thought that they were the countrymen of Washington; "he was very

---

Mason Locke Weems, *The Life of George Washington; with Curious Anecdotes, Equally Honourable to Himself and Exemplary to his Young Countrymen,* 9th ed. (Philadelphia, 1809), 3–18.

well, general, when we left America." — *"Ah, gentlemen!"* rejoined he, *"Washington can never be otherwise than well: — The measure of his fame is full — Posterity shall talk of him with reverence as the founder of a great empire, when my name shall be lost in the vortex of Revolutions!"*

Who then that has a spark of virtuous curiosity, but must wish to know the history of him whose name could thus awaken the sigh even of Bonaparte? But is not his history *already* known? Have not a thousand orators spread his fame abroad, bright as his own Potomac, when he reflects the morning sun, and flames like a sea of liquid gold, the wonder and delight of all the neighbouring shores? Yes, they have indeed spread his fame abroad. . . . his fame as Generalissimo of the armies, and first President of the councils of his nation. But this is not *half* his fame. . . . True, he is there seen in *greatness,* but it is only the greatness of public character, which is no evidence of *true greatness;* for a public character is often an artificial one. At the head of an army or nation, where gold and glory are at stake, and where a man feels himself the *burning focus* of unnumbered eyes; he must be a paltry fellow indeed, who does not play his part pretty handsomely . . . even the common passions of pride, avarice, or ambition, will put him up to his metal, and call forth his best and bravest doings. But let all this heat and blaze of public situation and incitement be withdrawn; let him be thrust back into the shade of private life, and you shall see how soon, like a forced plant robbed of its hot-bed, he will drop his false foliage and fruit, and stand forth confessed in native stickweed sterility and worthlessness. . . . There was Benedict Arnold — while strutting a BRIGADIER GENERAL on the public stage, he could play you the *great man,* on a handsome scale . . . he out-marched Hannibal, and out-fought Burgoyne . . . he chaced the British like curlews, or cooped them up like chickens! and yet in the *private walks of life,* in Philadelphia, he could swindle rum from the commissary's stores, and, with the aid of loose women, retail it by the gill!! . . . And there was the great duke of Marlborough too — his public character, a thunderbolt in war! Britain's boast, and terror of the French! But his private character, what? Why a *swindler* to whom not *Arnold's self* could hold a candle; a perfect nondescript of baseness; a shaver of farthings from the poor sixpenny pay of his own brave soldiers!!!

It is not then in the glare of *public,* but in the shade of *private life,* that we are to look for the man. Private life is always *real* life. Behind the curtain, where the eyes of the million are not upon him, and where a man can have no motive but *inclination,* no excitement but

honest nature, there he will always be sure to act *himself;* consequently, if he act greatly, he must be great indeed. Hence it has been
justly said, that, "our *private deeds,* if *noble,* are noblest of our lives."
   Of these private deeds of Washington very little has been said. In
most of the elegant orations pronounced to his praise, you see nothing
of Washington below *the clouds*—nothing of Washington the *dutiful
son*—the affectionate brother—the cheerful school-boy—the diligent
surveyor—the neat draftsman—the laborious farmer—and widow's
husband—the orphan's father—the poor man's friend. No! this is not
the Washington you see; 'tis only Washington the HERO, and the
Demigod.. . . . Washington the *sun beam* in council, or the *storm* in war.
   And in all the ensigns of character, amidst which he is generally
drawn, you see none that represent him what he really was, *"the
Jupiter Conservator,"* the *friend and benefactor of men.* Where's his
bright ploughshare that he loved—or his wheat-crowned fields, waving in yellow ridges before the wanton breeze—or his hills whitened
over with flocks—or his clover-covered pastures spread with innu-
merous herds—or his neat-clad servants, with songs rolling the
heavy harvest before them? Such were the scenes of *peace, plenty,* and
happiness, in which Washington delighted. But his eulogists have
denied him *these,* the only scenes which belong to man the GREAT, and
have trick'd him up in the vile drapery of man the *little.* See! there he
stands! with the port of Mars *"the destroyer,"* dark frowning over the
fields of war . . . the lightning of Potter's blade is by his side—the deep-
mouthed cannon is before him, disgorging its flesh-mangling balls—
his war-horse paws with impatience to bear him, a speedy thunderbolt,
against the pale and bleeding ranks of Britain!—These are the drawings usually given of Washington; drawings masterly no doubt, and
perhaps justly descriptive of him in some scenes of his life; but scenes
they were, which I am sure his soul *abhorred,* and in which at any rate,
you see nothing of his *private virtues.* These old fashioned commodi-
ties are generally thrown into the back ground of the picture, and
treated, as the grandees at the London and Paris routs, treat their
good old *aunts* and *grandmothers,* huddling them together into the
back rooms, there to wheeze and cough by themselves, and not
depress the fine laudanum-raised spirits of the *young sparklers.* And
yet it was to those *old-fashioned virtues* that our hero owed every
thing. For they in fact were the food of the great actions of him, whom
men call Washington. It was they that enabled him, first to triumph
over *himself,* then over the *British,* and uniformly to set such bright

examples of *human perfectibility* and *true greatness,* that compared therewith, the history of his capturing Cornwallis and Tarleton, with their buccaneering legions, sounds almost as *small* as the story of old General Putnam's catching his wolf and her lamb-killing whelps.[2]

Since then it is the private virtues that lay the foundation of all human excellence—since it was these that exalted Washington to be *"Columbia's first* and *greatest Son,"* be it our first care to present these, in all their lustre, before the admiring eyes of our *children.* To *them* his private character is *every thing;* his public, hardly *any thing.* For how glorious soever it may have been in Washington to have undertaken the emancipation of his country; to have stemmed the long tide of adversity; to have baffled every effort of a wealthy and warlike nation; to have obtained for his countrymen the completest victory, and for himself the most unbounded power; and then to have returned that power, accompanied with all the weight of his own great character and advice to establish a government that should immortalize the blessings of liberty . . . however glorious, I say, all this may have been to himself, or instructive to future generals and presidents, yet does it but *little* concern our *children.* For who among us can hope that his son shall ever be called, like Washington, to direct the storm of war, or to ravish the ears of deeply listening Senates? To be constantly placing him then, before our children, in this high character, what is it but like springing in the clouds a golden Phoenix, which no mortal calibre can ever hope to reach.[3] Or like setting pictures of the Mammoth before the *mice* whom "not all the manna of Heaven" can ever raise to equality?[4] Oh no! give us his *private virtues!* In *these,* every youth is interested, because in these every youth may become a Washington—a Washington in piety and patriotism,—in industry and honour—and consequently a Washington, in what alone deserves the name, SELF ESTEEM and UNIVERSAL RESPECT.

[2]General Israel Putnam (1718–1790), a New England folk hero, was known to contemporaries as "Old Wolf Put." When a large timber wolf killed 70 sheep on his Connecticut farm in 1742, Putnam tracked it to its den and crawled into a narrow cave to shoot it. Years later, as field commander of the Continental army at Bunker Hill, Putnam gave the famous order, "Men, you are marksmen—don't one of you fire until you see the white of their eyes."

[3]*phoenix:* unique person or thing

[4]The mammoth, a large extinct elephant, was believed to be the largest animal native to America.

CHAPTER II: BIRTH AND EDUCATION

Children like tender osiers take the bow;
And as they first are form'd for ever grow.

To this day numbers of good Christians can hardly find faith to believe that Washington was, bona fide, *a Virginian! "What! a buckskin!"* say they with a smile, *"George Washington a buckskin! pshaw! impossible! he was certainly an European: So great a man could never have been born in America."*

So great a man could never have been born in America!—Why that's the very *prince of reasons* why he should have been born here! Nature, we know, is fond of *harmonies;* and *paria paribus,* that is, *great things to great,* is the rule she delights to work by. Where, for example, do we look for the *whale* "the biggest born of nature?" not, I trow,[5] in a *millpond,* but in the main ocean; *"there go the great ships,"* and there are the spoutings of whales amidst their boiling foam.

By the same rule, where shall we look for Washington, the greatest among men, but in *America?* That greatest Continent, which, rising from beneath the frozen pole, stretches far and wide to the south, running almost *"whole the length of this vast terrene,"* and sustaining on her ample sides the roaring shock of half the watery globe. And equal to its size, is the furniture of this vast continent, where the Almighty has reared his cloud-capt mountains, and spread his sea-like lakes, and poured his mighty rivers, and hurled down his thundering cataracts in a style of the *sublime,* so far superior to any thing of the kind in the other continents, that we may fairly conclude that great men and great deeds are designed for America.

This seems to be the verdict of honest analogy; and accordingly we find America the honoured cradle of Washington, who was born on Pope's creek, in Westmoreland county, Virginia, the 22d of February, 1732. His father, whose name was Augustin Washington, was also a Virginian, but his grandfather (John) was an Englishman, who came over and settled in Virginia in 1657.

His father fully persuaded that a marriage of virtuous love comes nearest to angelic life, early stepped up to the *altar* with glowing cheeks and joy sparkling eyes, while by his side, with soft warm hand, sweetly trembling in his, stood the angel form of the lovely Miss Dandridge.

[5] *trow:* think, believe

After several years of great domestic happiness, Mr. Washington was separated, by death, from this excellent woman, who left him and two children to lament her early fate.

Fully persuaded still, that *"it is not good for man to be alone,"* he renewed, for the second time, the chaste delights of matrimonial love. His consort was Miss Mary Ball, a young lady of fortune, and descended from one of the best families in Virginia.

From his intermarriage with this charming girl, it would appear that our Hero's father must have possessed either a very pleasing person, or highly polished manners, or perhaps *both; for,* from what I can learn, he was at that time at least 40 years old! while she, on the other hand, was universally toasted as the belle of the Northern Neck,[6] and in the full bloom and freshness of love-inspiring sixteen. This I have from one who tells me that he has carried down many a sett dance with her; I mean that amiable and pleasant old gentleman, John Fitzhugh, Esq. of Stafford, *who* was, all his life, a neighbour and intimate of the Washington family. By his first wife, Mr. Washington had two children, both sons—Lawrence and Augustin. By his second wife, he had five children, four sons and a daughter—George, Samuel, John, Charles, and Elizabeth. Those *over delicate* ones, who are ready to faint at thought of a second marriage, might do well to remember, that the greatest man that ever lived was the son of this second marriage!

Little George had scarcely attained his fifth year, when his father left Pope's creek, and came up to a plantation which he had in Stafford, opposite to Fredericksburg. The house in which he lived is still to be seen. It lifts its low and modest front of faded red, over the turbid waters of Rappahannock; whither, to this day, numbers of people repair, and, with emotions unutterable, looking at the weatherbeaten mansion, exclaim, *"Here's the house where the Great Washington was born!"*

But it is all a mistake; for he was born, as I said, at Pope's creek, in Westmoreland county, near the margin of his own roaring Potomac.

The first place of education to which George was ever sent, was a little *"old field school,"* kept by one of his father's tenants, named Hobby; an honest, poor old man, who acted in the double character of sexton and schoolmaster. On his skill as a gravedigger, tradition is silent; but for a teacher of youth, his qualifications were certainly of the humbler sort; making what is generally called an A. B. C. schoolmaster. Such was the preceptor who first taught Washington the

---

[6]*Northern Neck:* area of Virginia between the Potomac and Rappahannock rivers

knowledge of letters! Hobby lived to see his young pupil in all his glory, and rejoiced exceedingly. In his cups—for, though a *sexton,* he would sometimes drink, particularly on the General's birth-days—he used to boast, that *"'twas he, who, between his knees, had laid the foundation of George Washington's greatness."*

But though George was early sent to a schoolmaster, yet he was not on that account neglected by his father. Deeply sensible of the *loveliness* and *worth* of which human nature is capable, through the *virtues* and *graces* early implanted in the heart, he never for a moment, lost sight of George in those all-important respects.

To assist his son to overcome that selfish spirit which too often leads children to fret and fight about trifles, was a notable care of Mr. Washington. For this purpose, of all the presents, such as cakes, fruit, &c. he received, he was always desired to give a liberal part to his play-mates. To enable him to do this with more alacrity, his father would remind him of the love which he would hereby gain, and the frequent presents which would in return be made *to him;* and also would tell of that great and good God, who delights above all things to see children love one another, and will assuredly reward them for acting so amiable a part.

Some idea of Mr. Washington's plan of education in this respect, may be collected from the following anecdote, related to me twenty years ago by an aged lady, who was a distant relative, and when a girl spent much of her time in the family.

"On a fine morning," said she, *"in the fall of 1737, Mr. Washington, having little George by the hand, came to the door and asked my cousin Washington and myself to walk with him to the orchard, promising he would show us a fine sight. On arriving at the orchard, we were presented with a fine sight indeed. The whole earth, as far as we could see, was strewed with fruit: and yet the trees were bending under the weight of apples, which hung in clusters like grapes, and vainly strove to hide their blushing cheeks behind the green leaves. Now, George, said his father, look here, my son! don't you remember when this good cousin of yours brought you that fine large apple last spring, how hardly I could prevail on you to divide with your brothers and sisters; though I promised you that if you would but do it, God Almighty would give you plenty of apples this fall. Poor George could not say a word; but hanging down his head, looked quite confused, while with his little naked toes he scratched in the soft ground. Now look up, my son, continued his father, look up, George! and see there how richly the blessed God has made good my promise to you. Wherever you turn your eyes, you see the trees loaded*

with fine fruit; many of them indeed breaking down, while the ground is covered with mellow apples more than you could ever eat, my son, in all your life time."

George looked in silence on the wide wilderness of fruit; he marked the busy humming bees, and heard the gay notes of birds, then lifting his eyes filled with shining moisture, to his father, he softly said, *"Well, Pa, only forgive me this time; see if I ever be so stingy any more."*

Some, when they look up to the oak whose giant arms throw a darkening shade over distant acres, or whose single trunk lays the keel of a man of war, cannot bear to hear of the time when this mighty plant was but an acorn, which a pig could have demolished: but others, who know their value, like to learn the soil and situation which best produces such noble trees. Thus, parents that are *wise* will listen well pleased, while I relate how moved the steps of the youthful Washington, whose single worth far outweighs all the oaks of Bashan and the red spicy cedars of Lebanon. Yes, they will listen delighted while I tell of their Washington in the days of his youth, when his little feet were swift towards the nests of birds; or when, wearied in the chace of the butterfly, he laid him down on his grassy couch and slept, while ministering spirits, with their roseate wings, fanned his glowing cheeks, and kissed his lips of innocence with that fervent love which makes *the Heaven!*

Never did the wise Ulysses take more pains with his beloved Telemachus, than did Mr. Washington with George, to inspire him with an *early love of truth.* "Truth, George," (said he) "is the loveliest quality of youth. I would ride fifty miles, my son, to see the little boy whose heart is so *honest* and his lips so *pure,* that we may depend on every word he says. O how lovely does such a child appear in the eyes of every body! His parents doat [dote] on him; his relations glory in him; they are constantly praising him to their children, whom they beg to imitate him. They are often sending for him, to visit them; and receive him, when he comes, with as much joy as if he were a little angel, come to set pretty examples to their children.

"But, Oh! how different, George, is the case with the boy who is so given to lying, that nobody can believe a word he says! He is looked at with aversion wherever he goes, and parents dread to see him come among their children. Oh, George! my son! rather than see you come to this pass, dear as you are to my heart, gladly would I assist to nail you up in your little coffin, and follow you to your grave. Hard, indeed, would it be to me to give up my son, whose little feet are always so ready to run about with me, and whose fondly looking eyes and sweet prattle make so large a part of my happiness: but still I would give him up, rather than see him a common liar."

"Pa, (said George very seriously) do I ever tell lies?"

"No, George, I *thank God* you do not, my son; and I rejoice in the hope you never will. At least, you shall never, from me, have cause to be guilty of so shameful a thing. Many parents, indeed, even compel their children to this vile practice, by barbarously beating them for every little fault; hence, on the next offence, the little terrified creature slips out a *lie!* just to escape the rod. But as to yourself, George, you know I have *always* told you, and now tell you again, that, whenever by accident you do any thing wrong, which must often be the case, as you are but a poor little boy yet, without *experience* or *knowledge,* never tell a falsehood to conceal it; but come *bravely* up, my son, like a *little man,* and tell me of it: and instead of beating you, George, I will but the more honour and love you for it, my dear."

This, you'll say, was sowing good seed!—Yes, it was: and the crop, thank God, was, as I believe it ever will be, where a man acts the true parent, that is, the *Guardian Angel,* by his child.

The following anecdote is a *case in point.* It is too valuable to be lost, and too true to be doubted; for it was communicated to me by the same excellent lady to whom I am indebted for the last.

"When George," said she, "was about six years old, he was made the wealthy master of a *hatchet!* of which, like most little boys, he was immoderately fond, and was constantly going about chopping every thing that came in his way. One day, in the garden, where he often amused himself hacking his mother's pea-sticks, he unluckily tried the edge of his hatchet on the body of a beautiful young English cherry-tree, which he barked so terribly, that I don't believe the tree ever got the better of it. The next morning the old gentleman finding out what had befallen his tree, which, by the by, was a great favourite, came into the house, and with much warmth asked for the mischievous author, declaring at the same time, that he would not have taken five guineas for his tree. Nobody could tell him any thing about it. Presently George and his hatchet made their appearance. *George,* said his father, *do you know who killed that beautiful little cherry-tree yonder in the garden?* This was a *tough question;* and George staggered under it for a moment; but quickly recovered himself: and looking at his father, with the sweet face of youth brightened with the inexpressible charm of all-conquering truth, he bravely cried out, *"I can't tell a lie, Pa, You know I can't tell a lie. I did cut it with my hatchet."*—*Run to my arms, you dearest boy,* cried his father in transports, *run to my arms; glad am I, George, that you killed my tree; for you have paid me for it a thousand fold. Such an act of heroism in my son, is more worth than a thousand trees, though blossomed with silver, and their fruits of purest gold.*

It was in this way, by interesting at once both his *heart* and *head,* that Mr. Washington conducted George with great ease and pleasure along the happy paths of virtue. But well knowing that his beloved charge, soon to be a man, would be left exposed to numberless temptations, both from himself and from others, his heart throbbed with the tenderest anxiety to make him acquainted with that GREAT BEING, whom to know and love, is to possess the surest defence against vice, and the best of all motives to virtue and happiness. To startle George into a lively sense of his Maker, he fell upon the following very curious but impressive expedient:

One day he went into the garden, and prepared a little bed of finely pulverized earth, on which he wrote George's name at full, in large letters—then strewing in plenty of cabbage seed, he covered them up and smoothed all over nicely with the roller. This bed he purposely prepared close along side of a gooseberry walk, which happening at this time to be well hung with ripe fruit, he knew would be honoured with George's visits pretty regularly every day. Not many mornings had passed away before in came George, with eyes wild rolling, and his little cheeks ready to burst with *great news.*

"O Pa! come here! come here!"

"What's the matter, my son, what's the matter?"

"O come here, I tell you, Pa, come here! and I'll show you such a sight as you never saw in all your life time."

The old gentleman suspecting what George would be at, gave him his hand, which he seized with great eagerness, and tugging him along through the garden, led him up point blank to the bed whereon was inscribed, in large letters, and in all the freshness of newly sprung plants, the full name of

GEORGE WASHINGTON

"There, Pa!" said George, quite in an ecstasy of astonishment, "did you ever see such a sight in all your life time?"

"Why it seems like a curious affair, sure enough, George!"

"But, Pa, who did make it there, who did make it there?"

"It grew there by *chance,* I suppose my son."

"By *chance,* Pa! O no! no! it never did grow there by *chance,* Pa; indeed that it never did!"

"High! why not, my son?"

"Why, Pa, did you ever see any body's name in a plant bed before?"

"Well, but George, such a thing might happen, though you never saw it before!"

"Yes, Pa, but I did never see the little plants grow up so as to make *one single* letter of my name before. Now, how could they grow up so as to make *all* the letters of my name! and then standing one after another, to spell *my name so exactly!*—and all so neat and even too, at top and bottom!! O Pa, you must not say *chance* did all this. Indeed *somebody* did it; and I dare say now, Pa, *you* did do it just to scare *me*, because I am your little boy."

His father smiled, and said, "Well George, you have guessed right—I indeed *did* it; but not to *scare* you, my son; but to learn you a great thing which I wish you to understand. I want, my son, to introduce you to your *true* Father."

"High, Pa, an't *you* my *true* father, that has loved me, and been so good to me always?"

"Yes, George, I am your father as the world calls it: and I love you very dearly too. But yet with all my love for you, George, I am but a poor good-for-nothing sort of a father in comparison of one you have."

"Aye! I know, well enough whom you mean, Pa. You mean God Almighty, don't you?"

"Yes, my son, I mean him indeed. *He* is your *true* Father, George."

"But, Pa, where is God Almighty? I did never *see* him yet."

"True, my son; but though you never *saw* him, yet he is always with you. You did not see me when ten days ago I made this little plant bed, where you see your name in such beautiful green letters; but though you did not *see* me here, yet you know I was here!!"

"Yes, Pa, that I do—I know you was here."

"Well then, and as my son could not believe that *chance* had made and put together so exactly the *letters* of his name, (though only sixteen) then how can he believe that *chance* could have made and put together all those millions and millions of things that are now so exactly fitted to his good? That my son may look at every thing around him, see! what fine eyes he has got! and a little pug nose to smell the sweet flowers! and pretty ears to hear sweet sound! and a lovely mouth for his bread and butter! and O, the little ivory teeth to cut it for him! and the dear little tongue to prattle with his father! and precious little hands and fingers to hold his playthings! and beautiful little feet for him to run about upon! and when my little rogue of a son is tired with running about, then the still night comes for him to lie down, and his mother sings, and the little crickets chirp him to sleep!

and as soon as he has slept enough, and jumps up fresh and strong as a little buck; there the sweet golden light is ready for him! When he looks down into the water, there he sees the beautiful silver fishes for him! and up in the *trees* there are the apples, and peaches, and *thousands* of sweet fruits for him! and *all, all around* him, wherever my dear boy looks, he sees every thing just to his *wants and wishes;*—the bubbling springs with cool sweet water for him to drink! and the wood to make him sparkling fires when he is cold! and beautiful horses for him to ride! and strong oxen to work for him! and the *good* cows to give him milk! and bees to make sweet honey for his sweeter mouth! and the little lambs, with snowy wool, for beautiful clothes for him! Now, these and all the *ten thousand thousand other good things* more than my son can ever think of, and all so exactly fitted to his *use* and *delight*. . . . Now how could chance ever have done all this for my little son? Oh George! . . . "

He would have gone on, but George, who had hung upon his father's words with looks and eyes of all-devouring attention, here broke out—

"Oh Pa, that's enough! that's enough! It can't be chance, indeed, it can't be chance, that made and gave me all these things."

"What was it then, do you think, my son?"

"Indeed, Pa, I don't know, unless it was *God Almighty!*"

"Yes, George, he it was, my son, and nobody else."

"Well, but Pa, (continued George) does God Almighty give me *every thing?* Don't you give me *some things*, Pa?"

"I give *you* something, indeed! Oh! how can I give you any thing, George! I, who have nothing on earth that I can call my own, no, not even the breath I draw!"

"High, Pa! isn't that great big house your house, and this garden, and the horses yonder, and oxen, and sheep, and trees, and every thing, isn't all yours, Pa?"

"Oh no! my son! no! Why you make me shrink into nothing, George, when you talk of all these belonging to *me*, who can't even make *a grain of sand!* Oh, how could I, my son, have given life to those great oxen and horses, when I can't give life even to a fly?—no! for if the poorest fly were killed, it is not your father, George, nor all the men in the world, that could ever make him alive again!"

At this, George fell into a profound silence, while his pensive looks showed that his youthful soul was labouring with some idea never felt before. Perhaps it was at that moment, that the good Spirit of God ingrafted on his heart that germ of *piety*, which filled his after life with so many of the precious fruits of *morality*.

16

JOHN TRUMBULL

*The Death of General Warren
at the Battle of Bunker's Hill*

1786

Son of the Revolutionary governor of Connecticut, John Trumbull (1756–1843) graduated from Harvard College and joined the Continental army in Boston in 1775. He served as an aide-de-camp to General George Washington and as adjutant to General Horatio Gates. In 1777, Trumbull left the army to paint, first at home, then in Boston, and finally in Paris and London, where he studied with the American expatriate Benjamin West after the war. His first in a series of American historical paintings was *The Death of General Warren at the Battle of Bunker's Hill,* completed in London in 1786. The small size of the canvas was designed to accommodate engravers and encourage print reproductions in the United States. The painting contributed to Trumbull's growing reputation as the painter of the American Revolution.

Yale University Art Gallery, Trumbull Collection.

JOHN TRUMBULL

The Declaration of Independence, Philadelphia, 4 July 1776

1787

John Trumbull completed his painting *The Declaration of Independence* in London in 1787. The painting includes a life portrait of John Adams, who stands in the central group with the other principal founders. Engraved and reproduced many times, Trumbull's *Declaration of Independence* was also copied by the artist himself onto the walls of the central rotunda of the U.S. Capitol in 1817. It became one of the most widely known visual representations of the American Revolution.
Yale University Art Gallery, Trumbull Collection.

CHARLES WILLSON PEALE

George Washington at the Battle of Princeton

1780–1781

A native of Maryland, Charles Willson Peale (1741–1827) was a saddle maker, silversmith, artist, inventor, political activist, and later the owner of America's first museum of art, natural history, and technology. He painted several life portraits of Washington in the 1770s, including *George Washington at the Battle of Princeton,* an official portrait commissioned by the Supreme Executive Council of Pennsylvania in January 1779. As an officer in the Pennsylvania militia, Peale had participated in the American military victories at Trenton and Princeton. Copied two dozen times by the artist himself, the painting enjoyed great popularity in the early Republic.

Yale University Art Gallery. Given by the Associates in Fine Arts and Mrs. Henry B. Loomis in memory of Henry Bradford Loomis, B.A. 1875.

GILBERT STUART

George Washington

1797

Gilbert Stuart (1755–1828), a native of Rhode Island who studied painting in London with Benjamin West, was regarded as the finest American portraitist of his day. He returned home from England expressly to paint George Washington's portrait in 1795–96. After completing two busts and the full-length portrait *George Washington* (1797), Stuart reproduced them more than one hundred times over the next thirty years. One of the first copies of the portrait was purchased by the U.S. government in 1800 to display in the White House. Considered a national treasure, it was the only painting saved by Dolley Madison as she fled before the British forces who set fire to the White House in August 1814 during the War of 1812.

The White House Collection, copyright White House Historical Association.

Contesting Popular Culture

20

SUSANNA HASWELL ROWSON

Charlotte. A Tale of Truth
1794

Susanna Haswell Rowson (c. 1762–1824), the daughter of a British naval officer and royal customs official, was born in England, raised in Massachusetts, and returned to England as part of a prisoner exchange in 1778. After joining a theater company in Philadelphia in 1793, she finally settled in Boston in 1796. A successful actress, playwright, novelist, and founder of a respected girls' academy in Boston, Rowson published her work under her own name. Her fourth novel, Charlotte. A Tale of Truth, *was written and published in London in 1791 and republished in Philadelphia in 1794. The most popular novel in America until the publication of Harriet Beecher Stowe's* Uncle Tom's Cabin *in 1852,* Charlotte *went through more than two hundred editions and sold about forty thousand copies in its first decade.*

PREFACE

For the perusal of the young and thoughtless of the fair sex, this Tale of Truth is designed; and I could wish my fair readers to consider it as not merely the effusion of Fancy, but as a reality. The circumstances on which I have founded this novel were related to me some little time since by an old lady who had personally known Charlotte, though she

Susanna Haswell Rowson, *Charlotte. A Tale of Truth* (Philadelphia, 1794), v–vi, 7–10, 32–36, 55–63, 89–94, 158–64.

concealed the real names of the characters, and likewise the place where the unfortunate scenes were acted: yet as it was impossible to offer a relation to the public in such an imperfect state, I have thrown over the whole a slight veil of fiction, and substituted names and places according to my own fancy. The principal characters in this little tale are now consigned to the silent tomb: it can therefore hurt the feelings of no one; and may, I flatter myself, be of service to some who are so unfortunate as to have neither friends to advise, or understanding to direct them, through the various and unexpected evils that attend a young and unprotected woman in her first entrance into life.

While the tear of compassion still trembled in my eye for the fate of the unhappy Charlotte, I may have children of my own, said I, to whom this recital may be of use, and if to your own children, said Benevolence, why not to the many daughters of Misfortune who, deprived of natural friends, or spoilt by a mistaken education, are thrown on an unfeeling world without the least power to defend themselves from the snares not only of the other sex, but from the more dangerous arts of the profligate of their own.

Sensible as I am that a novel writer, at a time when such a variety of works are ushered into the world under that name, stands but a poor chance for fame in the annals of literature, but conscious that I wrote with a mind anxious for the happiness of that sex whose morals and conduct have so powerful an influence on mankind in general; and convinced that I have not wrote a line that conveys a wrong idea to the head or a corrupt wish to the heart, I shall rest satisfied in the purity of my own intentions, and if I merit not applause, I feel that I dread not censure.

If the following tale should save one hapless fair one from the errors which ruined poor Charlotte, or rescue from impending misery the heart of one anxious parent, I shall feel a much higher gratification in reflecting on this trifling performance, than could possibly result from the applause which might attend the most elegant finished piece of literature whose tendency might deprave the heart or mislead the understanding.

CHAPTER I: A BOARDING SCHOOL

"Are you for a walk," said Montraville to his companion, as they arose from table; "are you for a walk? or shall we order the chaise and proceed to Portsmouth?" Belcour preferred the former; and they sauntered out to view the town, and to make remarks on the inhabitants, as they returned from church.

Montraville was a Lieutenant in the army: Belcour was his brother officer: they had been to take leave of their friends previous to their departure for America, and were now returning to Portsmouth, where the troops waited orders for embarkation. They had stopped at Chichester to dine; and knowing they had sufficient time to reach the place of destination before dark, and yet allow them a walk, had resolved, it being Sunday afternoon, to take a survey of the Chichester ladies as they returned from their devotions.

They had gratified their curiosity, and were preparing to return to the inn without honouring any of the belles with particular notice, when Madame Du Pont, at the head of her school, descended from the church. Such an assemblage of youth and innocence naturally attracted the young soldiers: they stopped; and, as the little cavalcade passed, almost involuntarily pulled off their hats. A tall, elegant girl looked at Montraville and blushed: he instantly recollected the features of Charlotte Temple, whom he had once seen and danced with at a ball at Portsmouth. At that time he thought on her only as a very lovely child, she being then only thirteen; but the improvement two years had made in her person, and the blush of recollection which suffused her cheeks as she passed, awakened in his bosom new and pleasing ideas. Vanity led him to think that pleasure at again beholding him might have occasioned the emotion he had witnessed, and the same vanity led him to wish to see her again.

"She is the sweetest girl in the world," said he, as he entered the inn. Belcour stared. "Did you not notice her?" continued Montraville: "she had on a blue bonnet, and with a pair of lovely eyes of the same colour, has contrived to make me feel devilish odd about the heart."

"Pho," said Belcour, "a musket ball from our friends, the Americans, may in less than two months make you feel worse."

"I never think of the future," replied Montraville; "but am determined to make the most of the present, and would willingly compound with any kind Familiar who would inform me who the girl is, and how I might be likely to obtain an interview."

But no kind Familiar at that time appearing, and the chaise which they had ordered, driving up to the door, Montraville and his companion were obliged to take leave of Chichester and its fair inhabitant, and proceed on their journey.

But Charlotte had made too great an impression on his mind to be easily eradicated: having therefore spent three whole days in thinking on her and in endeavouring to form some plan for seeing her, he determined to set off for Chichester, and trust to chance either to favour or frustrate his designs. Arriving at the verge of the town, he dismounted, and sending the servant forward with the horses, proceeded toward the place, where, in the midst of an extensive pleasure ground, stood the mansion which contained the lovely Charlotte Temple. Montraville leaned on a broken gate, and looked earnestly at the house. The wall which surrounded it was high, and perhaps the Argus's who guarded the Hesperian fruit within, were more watchful than those famed of old.[1]

"'Tis a romantic attempt," said he; "and should I even succeed in seeing and conversing with her, it can be productive of no good: I must of necessity leave England in a few days, and probably may never return; why then should I endeavour to engage the affections of this lovely girl, only to leave her a prey to a thousand inquietudes, of which at present she has no idea? I will return to Portsmouth and think no more about her."

The evening now was closed; a serene stillness reigned; and the chaste Queen of Night with her silver crescent faintly illuminated the hemisphere. The mind of Montraville was hushed into composure by the serenity of the surrounding objects. "I will think on her no more," said he, and turned with an intention to leave the place; but as he turned, he saw the gate which led to the pleasure grounds open, and two women come out, who walked arm-in-arm across the field.

"I will at least see who these are," said he. He overtook them, and giving them the compliments of the evening, begged leave to see them into the more frequented parts of the town: but how was he delighted, when, waiting for an answer, he discovered, under the concealment of a large bonnet, the face of Charlotte Temple.

He soon found means to ingratiate himself with her companion, who was a French teacher at the school, and, at parting, slipped a letter he had purposely written, into Charlotte's hand, and five guineas into that of Mademoiselle, who promised she would endeavour to bring her young charge into the field again the next evening. . . .

---

[1]*Argus:* watchful guardian. In Greek mythology, an Argus was a monster with one hundred eyes; Hesperides nymphs, aided by a dragon, guarded a garden of golden apples.

CHAPTER VI: AN INTRIGUING TEACHER

Madame Du Pont was a woman every way calculated to take the care of young ladies, had that care entirely devolved on herself; but it was impossible to attend the education of a numerous school without proper assistants; and those assistants were not always the kind of people whose conversation and morals were exactly such as parents of delicacy and refinement would wish a daughter to copy. Among the teachers at Madame Du Pont's school, was Mademoiselle La Rue, who added to a pleasing person and insinuating address, a liberal education and the manners of a gentlewoman. She was recommended to the school by a lady whose humanity overstepped the bounds of discretion: for though she knew Miss La Rue had eloped from a convent with a young officer, and on coming to England had lived with several different men in open defiance of all moral and religious duties; yet, finding her reduced to the most abject want, and believing the penitence which she professed to be sincere, she took her into her own family, and from thence recommended her to Madame Du Pont, as thinking the situation more suitable for a woman of her abilities. But Mademoiselle possessed too much of the spirit of intrigue to remain long without adventures. At church, where she constantly appeared, her person attracted the attention of a young man who was upon a visit at a gentleman's seat[2] in the neighbourhood: she had met him several times clandestinely; and being invited to come out that evening, and eat some fruit and pastry in a summer-house belonging to the gentleman he was visiting, and requested to bring some of the ladies with her, Charlotte being her favourite, was fixed on to accompany her.

The mind of youth eagerly catches at promised pleasure: pure and innocent by nature, it thinks not of the dangers lurking beneath those pleasures, till too late to avoid them: when Mademoiselle asked Charlotte to go with her, she mentioned the gentleman as a relation, and spoke in such high terms of the elegance of his gardens, the sprightliness of his conversation, and the liberality with which he ever entertained his guests, that Charlotte thought only of the pleasure she should enjoy in the visit, -not on the imprudence of going without her governess's knowledge, or of the danger to which she exposed herself in visiting the house of a gay young man of fashion.

---

[2]*gentleman's seat:* estate

Madame Du Pont was gone out for the evening, and the rest of the ladies retired to rest, when Charlotte and the teacher stole out at the back gate, and in crossing the field, were accosted by Montraville, as mentioned in the first chapter.

Charlotte was disappointed in the pleasure she had promised herself from this visit. The levity of the gentlemen and the freedom of their conversation disgusted her. She was astonished at the liberties Mademoiselle permitted them to take; grew thoughtful and uneasy, and heartily wished herself at home again in her own chamber.

Perhaps one cause of that wish might be, an earnest desire to see the contents of the letter which had been put into her hand by Montraville.

Any reader who has the least knowledge of the world, will easily imagine the letter was made up of encomiums on her beauty, and vows of everlasting love and constancy; nor will he be surprised that a heart open to every gentle, generous sentiment, should feel itself warmed by gratitude for a man who professed to feel so much for her; nor is it improbable but her mind might revert to the agreeable person and martial appearance of Montraville.

In affairs of love, a young heart is never in more danger than when attempted by a handsome young soldier. A man of an indifferent appearance, will, when arrayed in a military habit, shew to advantage; but when beauty of person, elegance of manner, and an easy method of paying compliments, are united to the scarlet coat, smart cockade, and military sash, ah! well-a-day for the poor girl who gazes on him: she is in imminent danger; but if she listens to him with pleasure, 'tis all over with her, and from that moment she has neither eyes nor ears for any other object.

Now, my dear sober matron, (if a sober matron should deign to turn over these pages, before she trusts them to the eye of a darling daughter,) let me intreat you not to put on a grave face, and throw down the book in a passion and declare 'tis enough to turn the heads of half the girls in England; I do solemnly protest, my dear madam, I mean no more by what I have here advanced, than to ridicule those romantic girls, who foolishly imagine a red coat and silver epaulet constitute the fine gentleman; and should that fine gentleman make half a dozen fine speeches to them, they will imagine themselves so much in love as to fancy it a meritorious action to jump out of a two pair of stairs[3] window, abandon their friends, and trust entirely to the honour

---

[3] *two pair of stairs:* second-story

of a man, who perhaps hardly knows the meaning of the word, and if he does, will be too much the modern man of refinement, to practice it in their favour.

Gracious heaven! when I think on the miseries that must rend the heart of a doating [doting] parent, when he sees the darling of his age at first seduced from his protection, and afterwards abandoned, by the very wretch whose promises of love decoyed her from the paternal roof—when he sees her poor and wretched, her bosom torn between remorse for her crime and love for her vile betrayer—when fancy paints to me the good old man stooping to raise the weeping penitent, while every tear from her eye is numbered by drops from his bleeding heart, my bosom glows with honest indignation, and I wish for power to extirpate those monsters of seduction from the earth.

Oh my dear girls—for to such only am I writing—listen not to the voice of love, unless sanctioned by paternal approbation: be assured, it is now past the days of romance: no woman can be run away with contrary to her own inclination: then kneel down each morning, and request kind heaven to keep you free from temptation, or, should it please to suffer you to be tried, pray for fortitude to resist the impulse of inclination when it runs counter to the precepts of religion and virtue. . . .

CHAPTER XI: CONFLICT OF LOVE AND DUTY

Almost a week was now gone, and Charlotte continued every evening to meet Montraville, and in her heart every meeting was resolved to be the last; but alas! when Montraville at parting would earnestly intreat one more interview, that treacherous heart betrayed her; and, forgetful of its resolution, pleaded the cause of the enemy so powerfully, that Charlotte was unable to resist. Another and another meeting succeeded; and so well did Montraville improve each opportunity, that the heedless girl at length confessed no idea could be so painful to her as that of never seeing him again.

"Then we will never be parted," said he.

"Ah, Montraville," replied Charlotte, forcing a smile, "how can it be avoided? My parents would never consent to our union; and even could they be brought to approve it, how should I bear to be separated from my kind, my beloved mother?"

"Then you love your parents more than you do me, Charlotte?"

"I hope I do," said she, blushing and looking down, "I hope my affection for them will ever keep me from infringing the laws of filial duty."

"Well, Charlotte," said Montraville gravely, and letting go her hand, "since that is the case, I find I have deceived myself with fallacious hopes. I had flattered my fond heart, that I was dearer to Charlotte than any thing in the world beside. I thought that you would for my sake have braved the dangers of the ocean, that you would, by your affection and smiles, have softened the hardships of war, and, had it been my fate to fall, that your tenderness would cheer the hour of death and smooth my passage to another world. But farewel[l], Charlotte! I see you never loved me. I shall now welcome the friendly ball that deprives me of the sense of my misery."

"Oh stay, unkind Montraville," cried she, catching hold of his arm, as he pretended to leave her, "stay, and to calm your fears, I will here protest that was it not for the fear of giving pain to the best of parents, and returning their kindness with ingratitude, I would follow you through every danger, and, in studying to promote your happiness, insure my own. But I cannot break my mother's heart, Montraville; I must not bring the grey hairs of my doating grand-father with sorrow to the grave, or make my beloved father perhaps curse the hour that gave me birth." She covered her face with her hands, and burst into tears.

"All these distressing scenes, my dear Charlotte," cried Montraville, "are merely the chimeras of a disturbed fancy. Your parents might perhaps grieve at first; but when they heard from your own hand that you was with a man of honour, and that it was to insure your felicity by an union with him, to which you feared they would never have given their assent, that you left their protection, they will, be assured, forgive an error which love alone occasioned, and when we return from America, receive you with open arms and tears of joy."

Belcour and Mademoiselle heard this last speech, and conceiving it a proper time to throw in their advice and persuasions, approached Charlotte, and so well seconded the intreaties of Montraville, that finding Mademoiselle intended going with Belcour, and feeling her own treacherous heart too much inclined to accompany them, the hapless Charlotte, in an evil hour, consented that the next evening they should bring a chaise to the end of the town, and that she would leave her friends, and throw herself entirely on the protection of Montraville. "But should you," said she, looking earnestly at him, her eyes full of tears, "should you, forgetful of your promises, and repenting the engagements you here voluntarily enter into, forsake and leave me on a foreign shore—"

"Judge not so meanly of me," said he. "The moment we reach our place of destination, Hymen shall sanctify our love; and when I shall forget your goodness, may heaven forget me."

"Ah," said Charlotte, leaning on Mademoiselle's arm as they walked up the garden together, "I have forgot all that I ought to have remembered, in consenting to this intended elopement."

"You are a strange girl," said Mademoiselle: "you never know your own mind two minutes at a time. Just now you declared Montraville's happiness was what you prized most in the world; and now I suppose you repent having insured that happiness by agreeing to accompany him abroad."

"Indeed I do repent," replied Charlotte, "from my soul: but while discretion points out the impropriety of my conduct, inclination urges me on to ruin."

"Ruin! fiddlestick!" said Mademoiselle; "am I not going with you? and do I feel any of these qualms?"

"You do not renounce a tender father and mother," said Charlotte.

"But I hazard my dear reputation," replied Mademoiselle, bridling.

"True," replied Charlotte, "but you do not feel what I do." She then bade her good night: but sleep was a stranger to her eyes, and the tear of anguish watered her pillow.

CHAPTER XII

Nature's last, best gift:
Creature in whom excell'd, whatever could
To sight or thought be nam'd!
Holy, divine! good, amiable, and sweet!
How thou art fall'n! —

When Charlotte left her restless bed, her languid eye and pale cheek discovered to Madame Du Pont the little repose she had tasted.

"My dear child," said the affectionate governess, "what is the cause of the languor so apparent in your frame? Are you not well?"

"Yes, my dear Madam, very well," replied Charlotte, attempting to smile, "but I know not how it was; I could not sleep last night, and my spirits are depressed this morning."

"Come chear up, my love," said the governess; "I believe I have brought a cordial to revive them. I have just received a letter from your good mama, and here is one for yourself."

Charlotte hastily took the letter: it contained these words—

"As to-morrow is the anniversary of the happy day that gave my beloved girl to the anxious wishes of a maternal heart, I have requested your governess to let you come home and spend it with us; and as I know you to be a good affectionate child, and make it your study to improve in those branches of education which you know will give most pleasure to your delighted parents, as a reward for your diligence and attention I have prepared an agreeable surprise for your reception. Your grand-father, eager to embrace the darling of his aged heart, will come in the chaise for you; so hold yourself in readiness to attend him by nine o'clock. Your dear father joins in every tender wish for your health and future felicity, which warms the heart of my dear Charlotte's affectionate mother,

L. TEMPLE."

"Gracious heaven!" cried Charlotte, forgetting where she was, and raising her streaming eyes as in earnest supplication.

Madame Du Pont was surprised. "Why these tears, my love?" said she. "Why this seeming agitation? I thought the letter would have rejoiced, instead of distressing you."

"It does rejoice me," replied Charlotte, endeavouring at composure, "but I was praying for merit to deserve the unremitted attentions of the best of parents."

"You do right," said Madame Du Pont, "to ask the assistance of heaven that you may continue to deserve their love. Continue, my dear Charlotte, in the course you have ever pursued, and you will insure at once their happiness and your own."

"Oh!" cried Charlotte, as her governess left her, "I have forfeited both for ever! Yet let me reflect:—the irrevocable step is not yet taken: it is not too late to recede from the brink of a precipice, from which I can only behold the dark abyss of ruin, shame, and remorse!"

She arose from her seat, and flew to the apartment of La Rue. "Oh Mademoiselle!" said she, "I am snatched by a miracle from destruction! This letter has saved me: it has opened my eyes to the folly I was so near committing. I will not go, Mademoiselle; I will not wound the hearts of those dear parents who make my happiness the whole study of their lives."

"Well," said Mademoiselle, "do as you please, Miss; but pray understand that my resolution is taken, and it is not in your power to alter it. I shall meet the gentlemen at the appointed hour, and shall not be surprized at any outrage which Montraville may commit, when he finds himself disappointed. Indeed I should not be astonished, was he to come immediately here, and reproach you for your instability in the hearing of the whole school: and what will be the consequence? you will bear the odium of having formed the resolution of eloping, and every girl of spirit will laugh at your want of fortitude to put it in execution, while prudes and fools will load you with reproach and contempt. You will have lost the confidence of your parents, incurred their anger, and the scoffs of the world; and what fruit do you expect to reap from this piece of heroism, (for such no doubt you think it is?) you will have the pleasure to reflect, that you have deceived the man who adores you, and whom in your heart you prefer to all other men, and that you are separated from him for ever."

This eloquent harangue was given with such volubility, that Charlotte could not find an opportunity to interrupt her, or to offer a single word till the whole was finished, and then found her ideas so confused, that she knew not what to say.

At length she determined that she would go with Mademoiselle to the place of assignation, convince Montraville of the necessity of adhering to the resolution of remaining behind; assure him of her affection, and bid him adieu.

Charlotte formed this plan in her mind, and exulted in the certainty of its success. "How shall I rejoice," said she, "in this triumph of reason over inclination, and, when in the arms of my affectionate parents, lift up my soul in gratitude to heaven as I look back on the dangers I have escaped!"

The hour of assignation arrived: Mademoiselle put what money and valuables she possessed in her pocket, and advised Charlotte to do the same; but she refused; "my resolution is fixed," said she; "I will sacrifice love to duty."

Mademoiselle smiled internally; and they proceeded softly down the back stairs and out of the garden gate. Montraville and Belcour were ready to receive them.

"Now," said Montraville, taking Charlotte in his arms, "you are mine for ever."

"No," said she, withdrawing from his embrace, "I am come to take an everlasting farewel[l]."

It would be useless to repeat the conversation that here ensued; suffice it to say, that Montraville used every argument that had formerly been successful, Charlotte's resolution began to waver, and he drew her almost imperceptibly towards the chaise.

"I cannot go," said she: "cease, dear Montraville, to persuade. I must not: religion, duty, forbid."

"Cruel Charlotte," said he, "if you disappoint my ardent hopes, by all that is sacred, this hand shall put a period to my existence. I cannot—will not live without you."

"Alas! my torn heart!" said Charlotte, "how shall I act?"

"Let me direct you," said Montraville, lifting her into the chaise.

"Oh! my dear forsaken parents!" cried Charlotte.

The chaise drove off. She shrieked, and fainted into the arms of her betrayer. . . .

CHAPTER XVIII: REFLECTIONS

"And am I indeed fallen so low," said Charlotte, "as to be only pitied? Will the voice of approbation no more meet my ear? and shall I never again possess a friend, whose face will wear a smile of joy whenever I approach? Alas! how thoughtless, how dreadfully imprudent have I been! I know not which is most painful to endure, the sneer of contempt, or the glance of compassion, which is depicted in the various countenances of my own sex: they are both equally humiliating. Ah! my dear parents, could you now see the child of your affections, the daughter whom you so dearly loved, a poor solitary being, without society, here wearing out her heavy hours in deep regret and anguish of heart, no kind friend of her own sex to whom she can unbosom her griefs, no beloved mother, no woman of character will appear in my company, and low as your Charlotte is fallen, she cannot associate with infamy."

These were the painful reflections which occupied the mind of Charlotte. Montraville had placed her in a small house a few miles from New-York: he gave her one female attendant, and supplied her with what money she wanted; but business and pleasure so entirely occupied his time, that he had little to devote to the woman, whom he had brought from all her connections, and robbed of innocence. Sometimes, indeed, he would steal out at the close of evening, and pass a few hours with her; and then so much was she attached to him, that all her sorrows were forgotten while blest with his society: she would enjoy a walk by moonlight, or sit by him in a little arbour at the bottom of the garden, and play on the harp, accompanying it with her plaintive, harmonious voice. But often, very often, did he promise to renew his visits, and, forgetful of his promise, leave her to mourn her disappointment. What painful hours of expectation would she pass! She would sit at a window which looked toward a field he used to cross, counting the minutes, and straining her eyes to catch the first glimpse of his person, till blinded with tears of disappointment, she would lean her head on her hands, and give free vent to her sorrows: then catching at some new hope, she would again renew her watchful position, till the shades of evening enveloped every object in a dusky cloud: she would then renew her complaints, and, with a heart bursting with disappointed love and wounded sensibility, retire to a bed which remorse had strewed with thorns, and court in vain that comforter of weary nature (who seldom visits the unhappy) to come and steep her senses in oblivion.

Who can form an adequate idea of the sorrow that preyed upon the mind of Charlotte? The wife, whose breast glows with affection to her husband, and who in return meets only indifference, can but faintly conceive her anguish. Dreadfully painful is the situation of such a woman, but she has many comforts of which our poor Charlotte was deprived. The duteous, faithful wife, though treated with indifference, has one solid pleasure within her own bosom, she can reflect that she has not deserved neglect—that she has ever fulfilled the duties of her station with the strictest exactness; she may hope, by constant assiduity and unremitted attention, to recall her wanderer, and be doubly happy in his returning affection; she knows he cannot leave her to unite himself to another: he cannot cast her out to poverty and contempt; she looks around her, and sees the smile of friendly welcome, or the tear of affectionate consolation, on the face of every person whom she favours with her esteem; and from all these circumstances she gathers comfort. But the poor girl by thoughtless passion led astray, who, in parting with her honour, has forfeited the esteem of the very man to whom she has sacrificed every thing dear and valuable in life, feels his indifference in the fruit of her own folly, and laments her want of power to recall his lost affection; she knows there is no tie but honour, and that, in a man who has been guilty of seduction, is but very feeble: he may leave her in a moment to shame and want; he may marry and forsake her for ever; and should he, she has no redress, no friendly, soothing companion to pour into her wounded mind the balm of consolation, no benevolent hand to lead her back to the path of rectitude; she has disgraced her friends, forfeited the good opinion of the world, and undone herself; she feels herself a poor solitary being in the midst of surrounding multitudes; shame bows her to the earth, remorse tears her distracted mind, and guilt, poverty, and disease close the dreadful scene: she sinks unnoticed to oblivion. The finger of contempt may point out to some passing daughter of youthful mirth, the humble bed where lies this frail sister of mortality; and will she, in the unbounded gaiety of her heart, exult in her own unblemished fame, and triumph over the silent ashes of the dead? Oh no! has she a heart of sensibility, she will stop, and thus address the unhappy victim of folly—

"Thou had'st thy faults, but sure thy sufferings have expiated them: thy errors brought thee to an early grave; but thou wert a fellow-creature—thou hast been unhappy—then be those errors forgotten."

Then, as she stoops to pluck the noxious weed from off the sod, a tear will fall, and consecrate the spot to Charity.

For ever honoured be the sacred drop of humanity; the angel of mercy shall record its source, and the soul from whence it sprang shall be immortal.

My dear Madam, contract not your brow into a frown of disapprobation. I mean not to extenuate the faults of those unhappy women who fall victims to guilt and folly; but surely, when we reflect how many errors we are ourselves subject to, how many secret faults lie hid in the recesses of our hearts, which we should blush to have brought into open day (and yet those faults require the lenity and pity of a benevolent judge, or awful would be our prospect of futurity) I say, my dear Madam, when we consider this, we surely may pity the faults of others.

Believe me, many an unfortunate female, who has once strayed into the thorny paths of vice, would gladly return to virtue, was any generous friend to endeavour to raise and re-assure her; but alas! it cannot be, you say; the world would deride and scoff. Then let me tell you, Madam, 'tis a very unfeeling world, and does not deserve half the blessings which a bountiful Providence showers upon it.

Oh, thou benevolent giver of all good! how shall we erring mortals dare to look up to thy mercy in the great day of retribution, if we now uncharitably refuse to overlook the errors, or alleviate the miseries, of our fellow-creatures. . . .

CHAPTER XXXIII: WHICH PEOPLE VOID OF FEELING NEED NOT READ

When Mrs. Beauchamp entered the apartment of the poor sufferer, she started back with horror. On a wretched bed, without hangings and but poorly supplied with covering, lay the emaciated figure of what still retained the semblance of a lovely woman, though sickness had so altered her features that Mrs. Beauchamp had not the least recollection of her person. In one corner of the room stood a woman washing, and, shivering over a small fire, two healthy but half naked children; the infant was asleep beside its mother, and, on a chair by the bed side, stood a porrenger and wooden spoon, containing a little gruel, and a tea-cup with about two spoonfulls of wine in it. Mrs. Beauchamp had never before beheld such a scene of poverty; she shuddered involuntarily, and exclaiming—"heaven preserve us!" leaned on the back of a chair ready to sink to the earth. The doctor repented having so precipitately brought her into this affecting scene; but there was no time for apologies: Charlotte caught the sound of her voice, and starting almost out of bed, exclaimed—"Angel of peace and mercy, art thou come to deliver me? Oh, I know you are, for whenever you was near me I felt eased of half my sorrows; but you don't know me, nor can I, with all the recollection I am mistress of, remember your name just now, but I know that benevolent countenance, and the softness of that voice which has so often comforted the wretched Charlotte."

Mrs. Beauchamp had, during the time Charlotte was speaking, seated herself on the bed and taken one of her hands; she looked at her attentively, and at the name of Charlotte she perfectly conceived the whole shocking affair. A faint sickness came over her. "Gracious heaven," said she, "is this possible?" and bursting into tears, she reclined the burning head of Charlotte on her own bosom; and folding her arms about her, wept over her in silence. "Oh," said Charlotte, "you are very good to weep thus for me: it is a long time since I shed a tear for myself: my head and heart are both on fire, but these tears of your's seem to cool and refresh it. Oh now I remember you said you would send a letter to my poor father: do you think he ever received it? or perhaps you have brought me an answer: why don't you speak, Madam? Does he say I may go home? Well he is very good; I shall soon be ready."

She then made an effort to get out of bed; but being prevented, her frenzy again returned, and she raved with the greatest wildness and incoherence. Mrs. Beauchamp, finding it was impossible for her to be removed, contented herself with ordering the apartment to be made more comfortable, and procuring a proper nurse for both mother and child; and having learnt the particulars of Charlotte's fruitless application to Mrs. Crayton from honest John,[4] she amply rewarded him for

---

[4]*Mrs. Crayton:* Mademoiselle La Rue's married name. John, Mrs. Crayton's servant, watched her deny knowing the pregnant and deserted Charlotte, and assisted Charlotte himself.

his benevolence, and returned home with a heart oppressed with many painful sensations, but yet rendered easy by the reflexion that she had performed her duty towards a distressed fellow-creature.

Early the next morning she again visited Charlotte, and found her tolerably composed; she called her by name, thanked her for her goodness, and when her child was brought to her, pressed it in her arms, wept over it, and called it the offspring of disobedience. Mrs. Beauchamp was delighted to see her so much amended, and began to hope she might recover, and, spite of her former errors, become an useful and respectable member of society; but the arrival of the doctor put an end to these delusive hopes: he said nature was making her last effort, and a few hours would most probably consign the unhappy girl to her kindred dust.

Being asked how she found herself, she replied—"Why better, much better, doctor. I hope now I have but little more to suffer. I had last night a few hours sleep, and when I awoke recovered the full power of recollection. I am quite sensible of my weakness; I feel I have but little longer to combat with the shafts of affliction. I have an humble confidence in the mercy of him who died to save the world, and trust that my sufferings in this state of mortality, joined to my unfeigned repentance, through his mercy, have blotted my offences from the sight of my offended maker. I have but one care—my poor infant! Father of mercy," continued she, raising her eyes, "of thy infinite goodness, grant that the sins of the parent be not visited on the unoffending child. May those who taught me to despise thy laws be forgiven; lay not my offences to their charge, I beseech thee; and oh! shower the choicest of thy blessings on those whose pity has soothed the afflicted heart, and made easy even the bed of pain and sickness."

She was exhausted by this fervent address to the throne of mercy, and though her lips still moved her voice became inarticulate: she lay for some time as it were in a doze, and then recovering, faintly pressed Mrs. Beauchamp's hand, and requested that a clergyman might be sent for.

On his arrival she joined fervently in the pious office,[5] frequently mentioning her ingratitude to her parents as what lay most heavy at her heart. When she had performed the last solemn duty, and was preparing to lie down, a little bustle on the outside door occasioned Mrs. Beauchamp to open it, and enquire the cause. A man in appearance about forty, presented himself, and asked for Mrs. Beauchamp.

---

[5]*pious office:* religious rite

"That is my name, Sir," said she.

"Oh then, my dear Madam," cried he, "tell me where I may find my poor, ruined, but repentant child."

Mrs. Beauchamp was surprised and affected; she knew not what to say; she foresaw the agony this interview would occasion Mr. Temple, who had just arrived in search of his Charlotte, and yet was sensible that the pardon and blessing of her father would soften even the agonies of death to the daughter.

She hesitated. "Tell me, Madam," cried he wildly, "tell me, I beseech thee, does she live? shall I see my darling once again? Perhaps she is in this house. Lead, lead me to her, that I may bless her, and then lie down and die."

The ardent manner in which he uttered these words occasioned him to raise his voice. It caught the ear of Charlotte: she knew the beloved sound: and uttering a loud shriek, she sprang forward as Mr. Temple entered the room. "My adored father." "My long lost child." Nature could support no more, and they both sunk lifeless into the arms of the attendants.

Charlotte was again put into bed, and a few moments restored Mr. Temple: but to describe the agony of his sufferings is past the power of any one, who, though they may readily conceive, cannot delineate the dreadful scene. Every eye gave testimony of what each heart felt—but all were silent.

When Charlotte recovered, she found herself supported in her father's arms. She cast on him a most expressive look, but was unable to speak. A reviving cordial was administered. She then asked, in a low voice, for her child: it was brought to her: she put it in her father's arms. "Protect her," said she, "and bless your dying—"

Unable to finish the sentence, she sunk back on her pillow: her countenance was serenely composed; she regarded her father as he pressed the infant to his breast with a steadfast look; a sudden beam of joy passed across her languid features, she raised her eyes to heaven—and then closed them for ever.

21

HUGH HENRY BRACKENRIDGE

*Modern Chivalry: Containing
the Adventures of Captain John Farrago,
and Teague O'Regan, His Servant*

1792

*Following his graduation from the College of New Jersey and brief stints
as a schoolmaster and chaplain to the Continental army, Hugh Henry
Brackenridge (1748–1816) moved in 1781 to the frontier town of Pittsburgh. There he practiced law, founded a newspaper and a boys' academy, and served in the Pennsylvania assembly until he was defeated in
his bid to represent his district in the state's ratifying convention for the
U.S. Constitution. Class and ethnic tensions were as important as political ideology in his defeat, as his Irish opponent (an ex-weaver and
Antifederalist) tarred the Scottish-born Brackenridge as an eastern elitist
and cosmopolitan Federalist.* Modern Chivalry: Containing the Adventures of Captain John Farrago, and Teague O'Regan, His Servant, *published in four volumes in Philadelphia and Pittsburgh between 1792 and
1797, formed part of Brackenridge's response. Although American critics
compared the work to Miguel de Cervantes's classic* Don Quixote, *it was
not widely read in the early Republic.*

CHAPTER I

John Farrago, was a man of about fifty-three years of age, of good natural sense, and considerable reading; but in some things whimsical,
owing perhaps to his greater knowledge of books than of the world;
but, in some degree, also, to his having never married, being what
they call an old batchelor, a characteristic of which is, usually, singu-
larity and whim. He had the advantage of having had in early life, an
academic education; but having never applied himself to any of the
learned professions, he had lived the greater part of his life on a small

Hugh Henry Brackenridge, *Modern Chivalry: Containing the Adventures of Captain John
Farrago, and Teague O'Regan, His Servant* (Philadelphia, 1792), 2 vols., 1, 11–16, 22–23,
25–36, 41–45.

farm, which he cultivated with servants or hired hands, as he could conveniently supply himself with either. The servant that he had at this time, was an Irishman, whose name was Teague O'Regan. I shall say nothing of the character of this man, because the very name imports what he was.[6]

A strange idea came into the head of Captain Farrago about this time; for, by the bye, I had forgot to mention that having being chosen captain of a company of militia in the neighborhood, he had gone by the name of Captain ever since; for the rule is, once a captain, and always a captain; but, as I was observing, the idea had come into his head, to saddle an old horse that he had, and ride about the world a little, with his man Teague at his heels, to see how things were going on here and there, and to observe human nature. For it is a mistake to suppose, that a man cannot learn man by reading him in a corner, as well as on the widest space of transaction. At any rate, it may yield amusement.

It was about a score of miles from his own house, that he fell in with what we call Races. The jockeys seeing him advance, with Teague by his side, whom they took for his groom, conceived him to be some person who had brought his horse to enter for the purse. Coming up and accosting him, said they, "You seem to be for the races, Sir; and have a horse to enter."

"Not at all," said the Captain; "this is but a common palfrey, and by no means remarkable for speed or bottom; he is a common plough horse which I have used on my farm for several years, and can scarce go beyond a trot; much less match himself with your blooded horses that are going to take the field on this occasion."

The jockeys were of opinion, from the speech, that the horse was what they call a bite, and that under the appearance of leanness and stiffness, there was concealed some hidden quality of swiftness uncommon. For they had heard of instances, where the most knowing had been taken in by mean looking horses; so that having laid two, or more, to one, they were nevertheless bit by the bet; and the mean looking nags, proved to be horses of a more than common speed and bottom. So that there is no trusting appearances. Such was the reasoning of the jockeys. For they could have no idea, that a man could come there in so singular a manner, with a groom at his foot, unless he had some great object of making money by the adventure. Under this idea, they began to interrogate him with respect to the blood and pedigree

---

[6]"Teague O'Regan" was a common epithet for an ignorant Irish immigrant.

of his horse: whether he was of the dove, or the bay mare that took the purse; and was imported by such a one at such a time? whether his sire was Tamerlane or Bajazet?

The Captain was irritated at the questions, and could not avoid answering—"Gentlemen," said he, "it is a strange thing that you should suppose that it is of any consequence what may be the pedigree of a horse. For even in men it is of no avail. Do we not find that sages have had blockheads for their sons; and that blockheads have had sages? It is remarkable, that as estates have seldom lasted three generations, so understanding and ability have seldom been transmitted to the second. There never was a greater man, take him as an orator and philosopher, than Cicero: and never was there a person who had greater opportunities than his son Marcus; and yet he proved of no account or reputation. This is an old instance, but there are a thousand others. Chesterfield and his son are mentioned. It is true, Philip and Alexander may be said to be exceptions: Philip of the strongest possible mind; capable of almost everything we can conceive; the deepest policy and the most determined valour; his son Alexander not deficient in the first, and before him in the last; if it is possible to be before a man than whom you can suppose nothing greater. It is possible, in modern times, that Tippo Saib may be equal to his father Hyder Ali.[7] Some talk of the two Pitts. I have no idea that the son is, in any respect, equal to old Sir William. The one is a laboured artificial minister: the other spoke with the thunder, and acted with the lightning of the gods. I will venture to say, that when the present John Adamses, and Lees, and Jeffersons, and Jays, and Henrys, and other great men, who figure upon the stage at this time, have gone to sleep with their fathers, it is an hundred to one if there is any of their descendants who can fill their places. Was I to lay a bet for a great man, I would sooner pick up the brat of a tinker, than go into the great houses to chuse a piece of stuff for a man of genius. Even with respect to personal appearance, which is more in the power of natural production, we do not see that beauty always produces beauty; but on the contrary, the homliest [homeliest] persons have oftentimes the best favored offspring; so that there is no rule or reason in these things.

"With respect to this horse, therefore, it can be of no moment whether he is blooded or studed [studded], or what he is. He is a good old horse, used to the plough, and carries my weight very well;

---

[7] *Tipu Sahib* (1751–1799), sultan of Mysore, India, at the end of the eighteenth century; *Hyder Ali* (1722–1782), Muslim ruler of Mysore, India

and I have never yet made enquiry with respect to his ancestors, or affronted him so much as to cast up to him the defect of parentage. I bought him some years ago from Niel Thomas, who had him from a colt. As far as I can understand, he was of a brown mare that John M'Neis had; but of what horse I know no more than the horse himself. His gaits are good enough, as to riding a short journey of seven or eight miles, or the like; but he is rather a pacer than a troter [trotter]; and though his bottom may be good enough in carrying a bag to the mill, or going in the plough, or the sled, or the harrow, etc., yet his wind is not so good, nor his speed, as to be fit for the heats."

The jockeys thought the man a fool, and gave themselves no more trouble about him. . . .

CHAPTER II: CONTAINING SOME GENERAL REFLECTIONS

The first reflection that arises, is, the good sense of the Captain, who was unwilling to impose his horse for a racer, not being qualified for the course. Because, as an old lean beast, attempting a trot, he was respectable enough; but going out of his nature, and affecting speed, he would have been contemptible. The great secret of preserving respect, is the cultivating and shewing to the best advantage the powers that we possess, and the not going beyond them. Everything in its element is good, and in their proper sphere all natures and capacities are excellent. This thought might be turned into a thousand different shapes, and cloathed with various expressions; but after all, it comes to the old proverb at last: *Ne sutor ultra crepidam,* Let the cobbler stick to his last; a sentiment we are about more to illustrate in the sequel of this work.

The second reflection that arises, is, the simplicity of the Captain, who was so unacquainted with the world, as to imagine that jockeys and men of the turf could be composed by reason and good sense; whereas there are no people who are by education of a less philosophic turn of mind. The company of horses is by no means favourable to good taste and genius. The rubbing and currying them, but little enlarges the faculties, or improves the mind; and even riding, by which a man is carried swiftly through the air, though it contributes to health, yet stores the mind with few or no ideas; and as men naturally consimilate with their company, so it is observable that your jockeys are a class of people not greatly removed from the sagacity of a good horse. Hence most probably the fable of the centaur, among the ancients; by which they held out the moral of the jockey and the horse being one beast. . . .

CHAPTER III

The Captain rising early next morning, and setting out on his way, had now arrived at a place where a number of people were convened, for the purpose of electing persons to represent them in the legislature of the state. There was a weaver who was a candidate for this appointment, and seemed to have a good deal of interest among the people. But another, who was a man of education, was his competitor. Relying on some talent of speaking which he thought he possessed, he addressed the multitude.

Said he, "Fellow citizens, I pretend not to any great abilities; but am conscious to myself that I have the best good will to serve you. But it is very astonishing to me, that this weaver should conceive himself qualified for trust. For though my acquirements are not great, yet his are still less. The mechanical business which he pursues, must necessarily take up so much of his time, that he cannot apply himself to political studies. I should therefore think it would be more answerable to your dignity, and conducive to your interest, to be represented by a man at least of some letters, than by an illiterate handicraftsman like this. It will be more honourable for himself, to remain at his loom and knot threads, than to come forward in a legislative capacity: because, in the one case, he is in the sphere where God and nature has placed him; in the other, he is like a fish out of water, and must struggle for breath in a new element.

"Is it possible he can understand the affairs of government, whose mind has been concentered to the small object of weaving webs; to the price by the yard, the grist of the thread, and such like matters as concern a manufacturer of cloths? The feet of him who weaves, are more occupied than the head, or at least as much; and therefore the whole man must be, at least, but in half accustomed to exercise his mental powers. For these reasons, all other things set aside, the chance is in my favour, with respect to information. However, you will decide, and give your suffrages to him or to me, as you shall judge expedient."

The Captain hearing these observations, and looking at the weaver, could not help advancing, and undertaking to subjoin something in support of what had been just said. Said he, "I have no prejudice against a weaver more than another man. Nor do I know any harm in the trade; save that from the sedentary life in a damp place, there is usually a paleness of the countenance: but this is a physical, not moral evil. Such usually occupy subterranean apartments; not for the purpose, like Demosthenes, of shaving their heads, and writing over eight times the history of Thucydides, and perfecting a stile of oratory,[8] but rather to keep the thread moist, or because this is considered but as an inglorious sort of trade, and is frequently thrust away into cellars, and damp outhouses, which are not occupied for a better use.

"But to rise from the cellar to the senate-house, would be an unnatural hoist. To come from counting threads, and adjusting them to the splits of a reed, to regulate the finances of a government, would be preposterous; there being no congruity in the case. There is no analogy between knotting threads and framing laws. It would be a reversion of the order of things. Not that a manufacturer of linen or woolen, or other stuff, is an inferior character, but a different one, from that which ought to be employed in affairs of state. It is unnecessary to enlarge on this subject; for you must all be convinced of the truth and propriety of what I say. But if you will give me leave to take the manufacturer aside a little, I think I can explain to him my ideas on the subject; and very probably prevail with him to withdraw his pretensions."

The people seeming to acquiesce, and beckoning to the weaver, they drew aside, and the Captain addressed him in the following words:

"Mr. Traddle," said he, for that was the name of the manufacturer, "I have not the smallest idea of wounding your sensibility; but it would seem to me, it would be more your interest to pursue your occupation, than to launch out into that of which you have no knowledge. When you go to the senate house, the application to you will not be to warp a web; but to make laws for the commonwealth. Now, suppose that the making these laws, requires a knowledge of commerce, or of the interests of agriculture, or those principles upon which the different manufacturers depend, what service could you render. It is possible you might think justly enough; but could you speak? You are not in the habit of public speaking. You are not furnished with those common-place ideas, with which even ignorant men can pass for knowing something. There is nothing makes a man so ridiculous as to attempt

---

Demosthenes (385?–322 B.C.), Athenian orator and statesman, perfected his style by copying the Greek historian Thucydides (471?–400 B.C.).

what is above his sphere. You are no tumbler for instance; yet should you give out that you could vault upon a man's back; or turn head over heels, like the wheel of a cart; the stiffness of your joints would encumber you; and you would fall upon your backside to the ground. Such a squash as that would do you damage. The getting up to ride on the state is an unsafe thing to those who are not accustomed to such horsemanship. It is a disagreeable thing for a man to be laughed at, and there is no way of keeping ones self from it but by avoiding all affectation."

While they were thus discoursing, a bustle had taken place among the crowd. Teague hearing so much about elections, and serving the government, took it into his head, that he could be a legislator himself. The thing was not displeasing to the people, who seemed to favour his pretensions, owing, in some degree, to there being several of his countrymen among the croud; but more especially to the fluctuation of the popular mind, and a disposition to what is new and ignoble. For though the weaver was not the most elevated object of choice, yet he was still preferable to this tatter-demalion, who was but a menial servant, and had so much of what is called the brogue on his tongue, as to fall far short of an elegant speaker.

The Captain coming up, and finding what was on the carpet, was greatly chagrined at not having been able to give the multitude a better idea of the importance of a legislative trust; alarmed also, from an apprehension of the loss of his servant. Under these impressions he resumed his address to the multitude. Said he, "This is making the matter still worse, gentlemen: this servant of mine is but a bog-trotter; who can scarcely speak the dialect in which your laws ought to be written; but certainly has never read a single treatise on any political subject; for the truth is, he cannot read at all. The young people of the lower class, in Ireland, have seldom the advantage of a good education; especially the descendants of the ancient Irish, who have most of them a great assurance of countenance, but little information, or literature. This young man, whose family name is O'Regan, has been my servant for several years. And, except a too great fondness for women, which now and then brings him into scrapes, he has demeaned himself in a manner tolerable enough. But he is totally ignorant of the great principles of legislation; and more especially, the particular interests of the government.

"A free government is a noble possession to a people: and this freedom consists in an equal right to make laws, and to have the benefit of the laws when made. Though doubtless, in such a government, the lowest citizen may become chief magistrate; yet it is sufficient to possess the right; not absolutely necessary to exercise it. Or even if you should think proper, now and then, to shew your privilege, and exert, in a signal manner, the democratic prerogative, yet is it not descending too low to filch away from me a hireling, which I cannot well spare, to serve your purposes? You are surely carrying the matter too far, in thinking to make a senator of this hostler;[9] to take him away from an employment to which he has been bred, and put him to another, to which he has served no apprenticeship: to set those hands which have been lately employed in currying my horse, to the draughting-bills, and preparing business for the house."

The people were tenacious of their choice, and insisted on giving Teague their suffrages; and by the frown upon their brows, seemed to indicate resentment at what had been said; as indirectly charging them with want of judgment; or calling in question their privilege to do what they thought proper.

"It is a very strange thing," said one of them, who was a speaker for the rest, "that after having conquered Burgoyne and Cornwallis, and got a government of our own, we cannot put in it whom we please. This young man may be your servant, or another man's servant; but if we chuse to make him a delegate, what is that to you. He may not be yet skilled in the matter, but there is a good day a-coming. We will impower him; and it is better to trust a plain man like him, than one of your high flyers, that will make laws to suit their own purposes."

Said the Captain, "I had much rather you would send the weaver, though I thought that improper, than to invade my household, and thus detract from me the very person that I have about to brush my boots, and clean my spurs." The prolocutor[10] of the people gave him to understand that his surmises were useless, for the people had determined on the choice, and Teague they would have for a representative.

Finding it answered no end to expostulate with the multitude, he requested to speak a word with Teague by himself. Stepping aside, he said to him, composing his voice, and addressing him in a soft manner, "Teague, you are quite wrong in this matter they have put into your head. Do you know what it is to be a member of a deliberative body? What qualifications are necessary? Do you understand any thing of geography? If a question should be, to make a law to dig a

---

[9] *hostler:* one who takes care of horses
[10] *prolocutor:* spokesman

canal in some part of the state, can you describe the bearing of the mountains, and the course of the rivers? Or if commerce is to be pushed to some new quarter, by the force of regulations, are you competent to decide in such a case? There will be questions of law, and astronomy on the carpet. How you must gape and stare like a fool, when you come to be asked your opinion on these subjects? Are you acquainted with the abstract principles of finance; with the funding [of] public securities; the ways and means of raising the revenue; providing for the discharge of the public debts, and all other things [with] respect [to] the economy of the government? Even if you had knowledge, have you a facility of speaking. I would suppose you would have too much pride to go to the house just to say, Ay, or No. This is not the fault of your nature, but of your education; having been accustomed to dig turf in your early years, rather than instructing yourself in the classics, or common school books.

"When a man becomes a member of a public body, he is like a raccoon, or other beast that climbs up the fork of a tree; the boys pushing at him with pitch-forks, or throwing stones, or shooting at him with an arrow, the dogs barking in the meantime. One will find fault with your not speaking; another with your speaking, if you speak at all. They will have you in the newspapers, and ridicule you as a perfect beast. There is what they call the caricatura; that is, representing you with a dog's head, or a cat's claw. As you have a red head, they will very probably make a fox of you, or a sorrel horse, or a brindled cow, or the like. It is the devil in hell to be exposed to the squibs and crackers of the gazette wits and publications.

"You know no more about these matters than a goose; and yet you would undertake rashly, without advice, to enter on the office; nay, contrary to advice. For I would not for a thousand guineas, though I have not the half of it to spare, that the breed of the O'Regans should come to this; bringing on them a worse stain than stealing sheep; to which they are addicted. You have nothing but your character, Teague, in a new country to depend upon. Let it never be said, that you quitted an honest livelihood, the taking care of my horse, to follow the newfangled whims of the times, and to be a statesman."

Teague was moved chiefly with the last part of the address, and consented to give up the object.

The Captain, glad of this, took him back to the people, and announced his disposition to decline the honor which they had intended him.

Teague acknowledged that he had changed his mind, and was willing to remain in a private station.

The people did not seem well pleased with the Captain; but as nothing more could be said about the matter, they turned their attention to the weaver, and gave him their suffrages. . . .

CHAPTER V: CONTAINING REFLECTIONS

A Democracy is beyond all question the freest government: because under this, every man is equally protected by the laws, and has equally a voice in making them. But I do not say an equal voice; because some men have stronger lungs than others, and can express more forcibly their opinions of public affairs. Others, though they may not speak very loud, yet have a faculty of saying more in a short time; and even in the case of others, who speak little or none at all, yet what they do say containing good sense, comes with greater weight; so that all things considered, every citizen, has not, in this sense of the word, an equal voice. But the right being equal, what great harm if it is unequally exercised? Is it necessary that every man should become a statesman? No more than that every man should become a poet or a painter. The sciences, are open to all; but let him only who has taste and genius pursue them. If any man covets the office of a bishop, says St. Paul, he covets a good work. But again, he adds this caution, Ordain not a novice, lest being lifted up with pride, he falls into the condemnation of the devil. It is indeed making a devil of a man to lift him up to a state to which he is not suited. A ditcher is a respectable character, with his over-alls on, and a spade in his hand; but put the same man to those offices which require the head, whereas he has been accustomed to impress with his foot, and there appears a contrast between the man and the occupation.

There are individuals in society, who prefer honour to wealth; or cultivate political studies as a branch of literary pursuits; and offer themselves to serve public bodies, in order to have an opportunity of discovering their knowledge, and exercising their judgment. It must be chagrining to these, and hurtful to the public, to see those who have no talent this way, and ought to have no taste, preposterously obtrude themselves upon the government. It is the same as if a bricklayer should usurp the office of a taylor, and come with his square and perpendicular, to take the measure of a pair of breeches.

It is proper that those who cultivate oratory, should go to the house of orators. But for an Ay and No man to be ambitious of that place, is to sacrifice his credit to his vanity.

I would not mean to insinuate that legislators are to be selected from the more wealthy of the citizens, yet a man's circumstances ought to be such as afford him leisure for study and reflection. There is often wealth without taste or talent. I have no idea, that because a man lives in a great house, and has a cluster of bricks or stones about his backside, that he is therefore fit for a legislator. There is so much pride and arrogance with those who consider themselves the first in a government, that it deserves to be checked by the populace, and the evil most usually commences on this side. Men associate with their own persons, the adventitious circumstances of birth and fortune: So that a fellow blowing with fat and repletion, conceives himself superior to the poor lean man, that lodges in an inferior mansion. But as in all cases, so in this, there is a medium. Genius and virtue are independent of rank and fortune; and it is neither the opulent, nor the indigent, but the man of ability and integrity that ought to be called forth to serve his country: and while, on the one hand, the aristocratic part of the government, arrogates a right to represent; on the other hand, the democratic contends the point; and from this conjunction and opposition of forces, there is produced a compound resolution, which carries the object in an intermediate direction. When we see, therefore, a Teague O'Regan lifted up, the philosopher will reflect, that it is to balance some purse-proud fellow, equally as ignorant, that comes down from the sphere of the aristocratic interest.

But every man ought to consider for himself, whether it is his use to be this draw-back, on either side. For as when good liquor is to be distilled, you throw in some material useless in itself to correct the effervescence of the spirit; so it may be his part to act as a sedative. For though we commend the effect, yet still the material retains but its original value.

But as the nature of things is such, let no man, who means well to the commonwealth, and offers to serve it, be hurt in his mind when someone of meaner talents is preferred. The people are a sovereign, and greatly despotic; but, in the main, just.

I have a great mind, in order to elevate the composition, to make quotations from the Greek and Roman history. And I am conscious to myself, that I have read over the writers on the government of Italy and Greece, in ancient, as well as modern times. But I have drawn a great deal more from reflection on the nature of things, than from all the writings I have ever read. Nay, the history of the election, which I have just given, will afford a better lesson to the American mind, than all that is to be found in other examples. We have seen here, a weaver a favoured candidate, and in the next instance, a bog-trotter superseding him. Now it may be said, that this is fiction; but fiction, or no fiction, the nature of the thing will make it a reality. But I return to the adventure of the Captain, whom I have upon my hands; and who, as far as I can yet discover, is a good honest man; and means what is benevolent and useful; though his ideas may not comport with the ordinary manner of thinking, in every particular.

22

ROYALL TYLER

The Contrast:
A Comedy in Five Acts
1787

A native of Boston, Royall Tyler (1757–1826) was a Harvard graduate, lawyer, army officer, and chief justice of the Vermont Supreme Court from 1807 to 1813. While in New York City on Massachusetts government business in March 1787, he saw an English comedy of manners, Richard Sheridan's School for Scandal, *performed at the John Street Theater. He decided that, with a few days' work, he could adapt the form for higher purposes. The* Contrast, *produced at the John Street Theater on April 16, 1787, became the first original play by an American citizen to be performed by a professional American theater company. The play also was staged in New York, Philadelphia, Baltimore, Boston, and Charleston and was printed in Philadelphia in 1790. Although the central plot contrasts the heroic Colonel Manly with the fashionable Billy Dimple, audiences preferred Manly's servant, Brother Jonathan, who became a stock character of American comedy.*

[Royall Tyler], *The Contrast* (Philadelphia, 1790), prologue, 39–52.

CHARACTERS

Col. Manly	Charlotte
Dimple	Maria
Van Rough	Letitia
Jessamy	Jenny
Jonathan	*Servants*

SCENE, *New York*

Prologue

*[Written by a Young Gentleman of New-York,
and Spoken by Mr. Wignell]*

Exult each patriot heart!—this night is shewn
A piece, which we may fairly call our own;
Where the proud titles of "My Lord! Your Grace!"
To humble Mr. and plain Sir give place.
Our Author pictures not from foreign climes
The fashions or the follies of the times;
But has confin'd the subject of his work
To the gay scenes—the circles of New-York.
On native themes his Muse displays her pow'rs;
If ours the faults, the virtues too are ours.

Why should our thoughts to distant countries roam,
When each refinement may be found at home?
Who travels now to ape the rich or great,
To deck an equipage[11] and roll in state;
To court the graces, or to dance with ease,
Or by hypocrisy to strive to please?
Our free-born ancestors such arts despis'd:
Genuine sincerity alone they priz'd;
Their minds, with honest emulation fir'd,
To solid good—not ornament—aspir'd;
Or, if ambition rous'd a bolder flame,
Stern virtue throve, where indolence was shame.

---

[11] *equipage:* horse-drawn carriage

But modern youths, with imitative sense,
Deem taste in dress the proof of excellence;
And spurn the meanness of your homespun arts,
Since homespun habits would obscure their parts;
Whilst all, which aims at splendour and parade,
Must come from Europe, and be ready made.
Strange! we should thus our native worth disclaim,

And check the progress of our rising fame.
Yet one, whilst imitation bears the sway,
Aspires to nobler heights, and points the way.
Be rous'd, my friends! his bold example view;
Let your own Bards be proud to copy you!
Should rigid critics reprobate our play,
At least the patriotic heart will say
"Glorious our fall, since in a noble cause.
"The bold attempt alone demands applause."

Still may the wisdom of the Comic Muse
Exalt your merits, or your faults accuse.
But think not, 'tis her aim to be severe; —
We all are mortals, and as mortals err.
If candour pleases, we are truly blest;
Vice trembles, when compell'd to stand confess'd.
Let not light Censure on your faults, offend,
Which aims not to expose them, but amend.
Thus does our Author to your candour trust;
Conscious, the free are generous, as just. . . . .

Act III Scene I

DIMPLE'S *Room*

. . .

Jessamy: Mrs. Jenny, I have the honour of presenting Mr. Jonathan, Colonel Manly's waiter, to you. I am extremely happy that I have it in my power to make two worthy people acquainted with each other's merits.

Jenny: So, Mr. Jonathan, I hear you were at the play last night.

Jonathan: At the play! why, did you think I went to the devil's drawing-room?

Jenny: The devil's drawing-room!

Jonathan: Yes; why an't cards and dice the devil's device, and the play-house the shop where the devil hangs out the vanities of the world upon the tenter-hooks of temptation? I believe you have not heard how they were acting the old boy one night, and the wicked one came among them sure enough, and went right off in a storm, and carried one quarter of the play-house with him. Oh! no, no, no! you won't catch me at a play-house, I warrant you.

Jenny: Well, Mr. Jonathan, though I don't scruple¹² your veracity, I have some reasons for believing you were there: pray, where were you about six o'clock?

Jonathan: Why, I went to see one Mr. Morrison, the *hocus pocus* man; they said as how he could eat a case knife.

Jenny: Well, and how did you find the place?

Jonathan: As I was going about here and there, to and again, to find it, I saw a great croud of folks going into a long entry that had lantherns over the door; so I asked a man whether that was not the place where they played *hocus pocus?* He was a very civil, kind man, though he did speak like the Hessians; he lifted up his eyes and said, "They play *hocus pocus* tricks enough there, Got [God] knows, mine friend."

Jenny: Well—

Jonathan: So I went right in, and they shewed me away, clean up to the garret, just like meeting-house gallery. And so I saw a power of topping folks, all sitting round in little cabins, "just like father's corn-cribs;" and then there was such a squeaking with the fiddles, and such a tarnal blaze with the lights, my head was near turned. At last the people that sat near me set up such a hissing—hiss— like so many mad cats; and then they went thump, thump, thump, just like our Peleg threshing wheat, and stamps away, just like the nation; and called out for one Mr. Langolee,—I suppose he helps act the tricks.

Jenny: Well, and what did you do all this time?

Jonathan: Gor, I—I liked the fun, and so I thumpt away, and hiss'd as lustily as the best of 'em. One sailor-looking man that sat by me, see-ing me stamp, and knowing I was a cute fellow, because I could make a roaring noise, clapt me on the shoulder and said, "You are a d——d hearty cock, smite my timbers!" I told him so I was, but I thought he need not swear so, and make use of such naughty words.

---

¹²*scruple:* doubt

Jenny: The savage!—Well, and did you see the man with his tricks?

Jonathan: Why, I vow, as I was looking out for him, they lifted up a great green cloth and let us look right into the next neighbor's house. Have you a good many houses in New-York made so in that 'ere way?

Jenny: Not many; but did you see the family?

Jonathan: Yes, swamp it; I see'd the family.

Jenny: Well, and how did you like them?

Jonathan: Why, I vow they were pretty much like other families;— there was a poor, good-natured, curse of a husband, and a sad rantipole[13] of a wife.

Jenny: But did you see no other folks?

Jonathan: Yes. There was one youngster; they called him Mr. Joseph; he talked as sober and as pious as a minister; but, like some ministers that I know, he was a sly tike in his heart for all that. He was going to ask a young woman to spark it with him, and—the Lord have mercy on my soul!—she was another man's wife.

Jessamy: The Wabash![14]

Jenny: And did you see any more folks?

Jonathan: Why, they came on as thick as mustard. For my part, I thought the house was haunted. There was a soldier fellow, who talked about his row de dow, dow, and courted a young woman; but, of all the cute folk I saw, I liked one little fellow—

Jenny: Aye! who was he?

Jonathan: Why, he had red hair, and a little round plump face like mine, only not altogether so handsome. His name was—Darby;— that was his baptizing name; his other name I forgot. Oh! it was Wig—Wag—Wag-all, Darby Wag-all,—pray, do you know him?—I should like to take a sling with him, or a drap [drop] of cider with a pepper-pod in it, to make it warm and comfortable.

Jenny: I can't say I have that pleasure.

Jonathan: I wish you did; he is a cute fellow. But there was one thing I didn't like in that Mr. Darby; and that was, he was afraid of some of them 'ere shooting irons, such as your troopers wear on training days. Now, I'm a true born Yankee American son of liberty, and I never was afraid of a gun yet in all my life.

Jenny: Well, Mr. Jonathan, you were certainly at the play-house.

Jonathan: I at the play-house!—Why didn't I see the play then?

---

[13]*rantipole:* scold, ranter
[14]*Wabash:* Indians who lived along the Wabash River, including the Miami

Jenny: Why, the people you saw were players.

Jonathan: Mercy on my soul! did I see the wicked players?—Mayhap that 'ere Darby that I liked so was the old serpent himself, and had his cloven foot in his pocket. Why, I vow, now I come to think on't, the candles seemed to burn blue, and I am sure where I sat it smelt tarnally of brimstone.

Jessamy: Well, Mr. Jonathan, from your account, which I confess is very accurate, you must have been at the play-house.

Jonathan: Why, I vow, I began to smell a rat. When I came away, I went to the man for my money again; you want your money? says he; yes, says I; for what? says he; why, says I, no man shall jocky me out of my money; I paid my money to see sights, and the dogs a bit of a sight have I seen, unless you call listening to people's private business a sight. Why, says he, it is the School for Scandalization.—The School for Scandalization!—Oh! ho! no wonder you New-York folks are so cute at it, when you go to school to learn it; and so I jogged off.

Jessamy: My dear Jenny, my master's business drags me from you; would to heaven I knew no other servitude than to your charms.

Jonathan: Well, but don't go; you won't leave me so—

Jessamy: Excuse me.—Remember the cash. *[Aside to him, and exits.]*

Jenny: Mr. Jonathan, won't you please to sit down? Mr. Jessamy tells me you wanted to have some conversation with me. *[Having brought forward two chairs, they sit.]*

Jonathan: Ma'am—

Jenny: Sir!—

Jonathan: Ma'am!—

Jenny: Pray, how do you like the city, Sir?

Jonathan: Ma'am!—

Jenny: I say, Sir, how do you like New-York?

Jonathan: Ma'am!—

Jenny: The stupid creature! but I must pass some little time with him, if it is only to endeavour to learn whether it was his master that made such an abrupt entrance into our house, and my young mistress's heart, this morning. *[Aside.]* As you don't seem to like to talk, Mr. Jonathan—do you sing?

Jonathan: Gor, I—I am glad she asked that, for I forgot what Mr. Jessamy bid me say, and I dare as well be hanged as act what he bid me do, I'm so ashamed. *[Aside.]* Yes, Ma'am, I can sing—I can sing Mear, Old Hundred, and Bangor.

Jenny: Oh! I don't mean psalm tunes. Have you no little song to please the ladies, such as Roslin Castle, or the Maid of the Mill?

Jonathan: Why, all my tunes [are] go to meeting tunes, save one, and I count you won't altogether like that 'ere.

Jenny: What is it called?

Jonathan:. I am sure you have heard folks talk about it; it is called Yankee Doodle.

Jenny: Oh! it is the tune I am fond of; and if I know anything of my mistress, she would be glad to dance to it. Pray, sing!

Jonathan: [Sings.]

Father and I went up to camp
Along with Captain Goodwin;
And there we saw the men and boys,
As thick as hasty-pudding.
       Yankee doodle do, etc.

And there we saw a swamping gun,
Big as log of maple,
On a little deuced cart
A load for father's cattle.
       Yankee doodle do, etc.

And every time they fired it off
It took a horn of powder
It made a noise—like father's gun,
Only a nation louder.
       Yankee doodle do, etc.

There was a man in our town,
His name was—

No, no, that won't do. Now, if I was with Tabitha Wymen and Jemima Cawley down at father Chase's, I shouldn't mind singing this all out before them—you would be affronted if I was to sing that, though that's a lucky thought; if you should be affronted, I have something dang'd cute, which Jessamy told me to say to you.

Jenny: Is that all! I assure you I like it of all things.

Jonathan: No, no; I can sing more; some other time, when you and I are better acquainted, I'll sing the whole of it—no, no—that's a fib—I can't sing but a hundred and ninety verses; our Tabitha at home can sing it all. *[Sings.]*

Marblehead's a rocky place,
And Cape-Cod is sandy;
Charlestown is burnt down,
Boston is the dandy.
       Yankee doodle, doodle do, etc.

I vow, my own town song has put me into such topping spirits that I believe I'll begin to do a little, as Jessamy says we must when we go a-courting. — *[Runs and kisses her.]* Burning rivers! cooling flames! red-hot roses! pig-nuts! hasty-pudding and ambrosia!

Jenny: What means this freedom? you insulting wretch. *[Strikes him.]*

Jonathan: Are you affronted?

Jenny: Affronted! with what looks shall I express my anger?

Jonathan: Looks! why as to the matter of looks, you look as cross as a witch.

Jenny: Have you no feeling for the delicacy of my sex?

Jonathan: Feeling! Gor, I—I feel the delicacy of your sex pretty smartly *[rubbing his cheek]*, though, I vow, I thought when you city ladies courted and married, and all that, you put feeling out of the question. But I want to know whether you are really affronted, or only pretend to be so? 'Cause, if you are certainly right down affronted, I am at the end of my tether; Jessamy didn't tell me what to say to you.

Jenny: Pretend to be affronted!

Jonathan: Aye, aye, if you only pretend, you shall hear how I'll go to work to make cherubim consequences. *[Runs up to her.]*

Jenny: Begone, you brute!

Jonathan: That looks like mad; but I won't lose my speech. My dearest Jenny—your name is Jenny, I think?—My dearest Jenny, though I have the highest esteem for the sweet favours you have just now granted me—Gor, that's a fib though, but Jessamy says it is not wicked to tell lies to the women. *[Aside.]* I say, though I have the highest esteem for the favours you have just now granted me, yet you will consider that, as soon as the dissolvable knot is tied, they will no longer be favours, but only matters of duty and matters of course.

Jenny: Marry you! you audacious monster! get out of my sight, or, rather, let me fly from you. *[Exit hastily.]*

Jonathan: Gor! she's gone off in a swinging passion, before I had time to think of consequences. If this is the way with your city ladies, give me the twenty acres of rock, the Bible, the cow, and Tabitha, and a little peaceable bundling.[15]

---

[15]Jonathan's expectations at home in New England. Bundling, a time-honored New England custom, allowed a man and woman, fully clothed, to share a bed while courting.

Scene 2

The Mall

[*Enter* MANLY.]

Manly: It must be so, Montague![16] and it is not all the tribe of Mande-villes[17] that shall convince me that a nation, to become great, must first become dissipated. Luxury is surely the bane of a nation: Luxury! which enervates both soul and body, by opening a thousand new sources of enjoyment, opens, also, a thousand new sources of contention and want: Luxury! which renders a people weak at home, and accessible to bribery, corruption, and force from abroad. When the Grecian states knew no other tools than the axe and the saw, the Grecians were a great, a free, and a happy people. The kings of Greece devoted their lives to the service of their country, and her senators knew no other superiority over their fellow-citizens than a glorious pre-eminence in danger and virtue. They exhibited to the world a noble spectacle,—a number of independent states united by a similarity of language, sentiment, manners, common interest, and common consent, in one grand mutual league of protection. And, thus united, long might they have continued the cherishers of arts and sciences, the protectors of the oppressed, the scourge of tyrants, and the safe asylum of liberty. But when foreign gold, and still more pernicious foreign luxury, had crept among them, they sapped the vitals of their virtue. The virtues of their ancestors were only found in their writings. Envy and suspicion, the vices of little minds, possessed them. The various states engendered jealousies of each other; and, more unfortunately, growing jealous of their great federal council, the Amphictyons, they forgot that their common safety had existed, and would exist, in giving them an honourable extensive prerogative. The common good was lost in the pursuit of private interest; and that people who, by uniting, might have stood against the world in arms, by dividing, crumbled into ruin;—their name is now only known in the page of the historian, and what they once were is all we have left to admire. Oh! that America! Oh! that my country, would, in this her day, learn the things which belong to her peace!

[16]Edward W. Montagu (1713–1776), English author of *Reflections on the Rise and Fall of the Antient Republicks*
[17]Bernard Mandeville (1670?–1733), English satirist, and author of *The Fable of the Bees, or Private Vices, Public Benefits*

[*Enter* DIMPLE.]

Dimple: You are Colonel Manly, I presume?

Manly: At your service, Sir.

Dimple: My name is Dimple, Sir. I have the honour to be a lodger in the same house with you, and, hearing you were in the Mall, came hither to take the liberty of joining you.

Manly: You are very obliging, Sir.

Dimple: As I understand you are a stranger here, Sir, I have taken the liberty to introduce myself to your acquaintance, as possibly I may have it in my power to point out some things in this city worthy your notice.

Manly: An attention to strangers is worthy a liberal mind, and must ever be gratefully received. But to a soldier, who has no fixed abode, such attentions are particularly pleasing.

Dimple: Sir, there is no character so respectable as that of a soldier. And, indeed, when we reflect how much we owe to those brave men who have suffered so much in the service of their country, and secured to us those inestimable blessings that we now enjoy, our liberty and independence, they demand every attention which gratitude can pay. For my own part, I never meet an officer, but I embrace him as my friend, nor a private in distress, but I insensibly extend my charity to him.—I have hit the Bumkin off very tolerably. [*Aside.*]

Manly: Give me your hand, Sir! I do not proffer this hand to everybody; but you steal into my heart. I hope I am as insensible to flattery as most men; but I declare (it may be my weak side) that I never hear the name of soldier mentioned with respect, but I experience a thrill of pleasure which I never feel on any other occasion.

Dimple: Will you give me leave, my dear Colonel, to confer an obligation on myself, by shewing you some civilities during your stay here, and giving a similar opportunity to some of my friends?

Manly: Sir, I thank you; but I believe my stay in this city will be very short.

Dimple: I can introduce you to some men of excellent sense, in whose company you will esteem yourself happy; and, by way of amusement, to some fine girls, who will listen to your soft things with pleasure.

Manly: Sir, I should be proud of the honour of being acquainted with those gentlemen:—but, as for the ladies, I don't understand you.

Dimple: Why, Sir, I need not tell you, that when a young gentleman is alone with a young lady he must say some soft things to her fair cheek—indeed, the lady will expect it. To be sure, there is not much pleasure when a man of the world and a finished coquette meet, who perfectly know each other; but how delicious is it to excite the emotions of joy, hope, expectation, and delight in the bosom of a lovely girl who believes every tittle of what you say to be serious!

Manly: Serious, Sir! In my opinion, the man who, under pretensions of marriage, can plant thorns in the bosom of an innocent, unsuspecting girl is more detestable than a common robber, in the same proportion as private violence is more despicable than open force, and money of less value than happiness.

Dimple: How he awes me by the superiority of his sentiments. *[Aside.]* As you say, Sir, a gentleman should be cautious how he mentions marriage.

Manly: Cautious, Sir! No person more approves of an intercourse between the sexes than I do. Female conversation softens our manners, whilst our discourse, from the superiority of our literary advantages, improves their minds. But, in our young country, where there is no such thing as gallantry, when a gentleman speaks of love to a lady, whether he mentions marriage or not, she ought to conclude either that he meant to insult her or that his intentions are the most serious and honourable. How mean, how cruel, is it, by a thousand tender assiduities, to win the affections of an amiable girl, and, though you leave her virtue unspotted, to betray her into the appearance of so many tender partialities, that every man of delicacy would suppress his inclination towards her, by supposing her heart engaged! Can any man, for the triv[i]al gratification of his leisure hours, affect the happiness of a whole life! His not having spoken of marriage may add to his perfidy, but can be no excuse for his conduct.

Dimple: Sir, I admire your sentiments;—they are mine. The light observations that fell from me were only a principle of the tongue; they came not from the heart; my practice has ever disapproved these principles.

Manly: I believe you, Sir. I should with reluctance suppose that those pernicious sentiments could find admittance into the heart of a gentleman.

Dimple: I am now, Sir, going to visit a family, where, if you please, I will have the honour of introducing you. Mr. Manly's ward, Miss Letitia, is a young lady of immense fortune; and his niece, Miss Charlotte Manly, is a young lady of great sprightliness and beauty.

Manly: That gentleman, Sir, is my uncle, and Miss Manly my sister.

Dimple: The devil she is! *[Aside.]* Miss Manly your sister, Sir? I rejoice to hear it, and feel a double pleasure in being known to you.— Plague on him! I wish he was at Boston again, with all my soul. *[Aside.]*

Manly: Come, Sir, will you go?

Dimple: I will follow you in a moment, Sir. *[Exit* MANLY.] Plague on it! this is unlucky. A fighting brother is a cursed appendage to a fine girl. Egad! I just stopped in time; had he not discovered himself, in two minutes more I should have told him how well I was with his sister. Indeed, I cannot see the satisfaction of an intrigue, if one can't have the pleasure of communicating it to our friends. *[Exit.]*

End of the Third Act

Encountering the Other

23

THOMAS JEFFERSON

Notes on the State of Virginia
1787

Written in response to a questionnaire about the American states circulated by the French legation in 1780, Thomas Jefferson's Notes on the State of Virginia *began as a statistical survey and expanded into the most comprehensive statement of his philosophical, social, political, and scientific beliefs. It was also his only published book. Jefferson (1743–1826) collected material for the book while serving as Virginia's Revolutionary governor, circulated the manuscript privately, and published it in Paris in 1784–85, in London in 1787, and in Philadelphia in 1788. Highly controversial, Jefferson's* Notes *became the subject of bitter partisan debate during the presidential campaigns of 1800 and 1804. Jefferson was labeled a "howling atheist," a "confirmed infidel," a "liar," and a "hypocrite." In 1798, he exclaimed, "O! that mine enemy would write a book! Has been a well known prayer against an enemy. I had written a book, and it has furnished matter of abuse for want of something better."*

Thomas Jefferson, *Notes on the State of Virginia* (Philadelphia, 1788), 40–41, 45, 61–66, 68–70, 99–100, 103–106, 145–50, 153–54.

QUERY VI: PRODUCTIONS MINERAL, VEGETABLE AND ANIMAL

... Our quadrupeds have been mostly described by Linnæus and Mons. de Buffon. Of these the Mammoth, or big buffalo, as called by the Indians, must certainly have been the largest. Their tradition is, that he was carnivorous, and still exists in the northern parts of America. A delegation of warriors from the Delaware tribe having visited the governor of Virginia, during the present revolution, on matters of business, after these had been discussed and settled in council, the governor asked them some questions relative to their country, and, among others, what they knew or had heard of the animal whose bones were found at the Saltlicks, on the Ohio. Their chief speaker immediately put himself into an attitude of oratory, and with a pomp suited to what he conceived the elevation of his subject, informed him that it was a tradition handed down from their fathers, "That in ancient times a herd of these tremendous animals came to the Big-bone licks, and began an universal destruction of the bear, deer, elks, buffaloes, and other animals, which had been created for the use of the Indians: that the Great Man above, looking down and seeing this, was so enraged that he seized his lightning, descended on the earth, seated himself on a neighbouring mountain, on a rock, of which his seat and the print of his feet are still to be seen, and hurled his bolts among them till the whole were slaughtered, except the big bull, who presenting his forehead to the shafts, shook them off as they fell; but missing one at length, it wounded him in the side; whereon, springing round, he bounded over the Ohio, over the Wabash, the Illinois, and finally over the great lakes, where he is living at this day." It is well known that on the Ohio, and in many parts of America further north, tusks, grinders, and skeletons of unparalleled magnitude, are found in great numbers, some lying on the surface of the earth, and some a little below it....

But to whatever animal we ascribe these remains, it is certain such a one has existed in America, and that it has been the largest of all terrestrial beings. It should have sufficed to have rescued the earth it inhabited, and the atmosphere it breathed, from the imputation of impotence in the conception and nourishment of animal life on a large scale: to have stifled, in its birth, the opinion of a writer, the most learned too of all others in the science of animal history, that in the new world, "La nature vivante est beaucoup moins agissante, beaucoup moins forte": that nature is less active, less energetic on one side of the globe than she is on the other. As if both sides were not warmed by the same genial sun; as if a soil of the same chemical composition, was less capable of elaboration into animal nutriment; as if the fruits and grains from that soil and sun, yielded a less rich chyle,[1] gave less extension to the solids and fluids of the body, or produced sooner in the cartilages, membranes, and fibres, that rigidity which restrains all further extension, and terminates animal growth. The truth is, that a Pigmy and a Patagonian, a Mouse and a Mammoth, derive their dimensions from the same nutritive juices. The difference of increment depends on circumstances unsearchable to beings with our capacities. Every race of animals seems to have received from their Maker certain laws of extension at the time of their formation. Their elaborative organs were formed to produce this, while proper obstacles were opposed to its further progress. Below these limits they cannot fall, nor rise above them. What intermediate station they shall take may depend on soil, on climate, on food, on a careful choice of breeders. But all the manna of heaven would never raise the mouse to the bulk of the mammoth. . . .

Hitherto I have considered this hypothesis as applied to brute animals only, and not in its extension to the man of America, whether aboriginal or transplanted. It is the opinion of Mons. de Buffon that the former furnishes no exception to it: "Although the savage of the new world is about the same height as man in our world, this does not suffice for him to constitute an exception to the general fact that all living nature has become smaller on that continent. The savage is feeble, and has small organs of generation; he has neither hair nor beard, and no ardor whatever for his female. Although swifter than the European because he is better accustomed to running, he is, on the other hand, less strong in body; he is also less sensitive, and yet more timid and cowardly. He has no vivacity, no activity of mind. The activity of his body is less an exercise, a voluntary motion, than a necessary action caused by want: relieve him of hunger and thirst, and you deprive him of the active principle of all his movements; he will rest stupidly upon his legs or lying down entire days. There is no need for seeking further the cause of the isolated mode of life of these savages and their repugnance for society: the most precious spark of the fire of nature has been refused to them; they lack ardor for their females, and consequently have no love for their fellow men. Not knowing this

[1] *chyle:* a milky fluid that aids in digestion

strongest and most tender of all affections, their other feelings are also cold and languid. They love their parents and children but little; the most intimate of all ties, the family connection, binds them therefore but loosely together. Between family and family there is no tie at all; hence they have no communion, no commonwealth, no state of society. Physical love constitutes their only morality; their heart is icy, their society cold, and their rule harsh. They look upon their wives only as servants for all work, or as beasts of burden, which they load without consideration with the burden of their hunting, and which they compel without mercy, without gratitude, to perform tasks which are often beyond their strength. They have only few children, and they take little care of them. Everywhere the original defect appears: they are indifferent because they have little sexual capacity, and this indifference to the other sex is the fundamental defect which weakens their nature, prevents its development, and—destroying the very germs of life—uproots society at the same time. Man is here no exception to the general rule. Nature, by refusing him the power of love, has treated him worse and lowered him deeper than any animal."[2] An afflicting picture indeed, which, for the honor of human nature, I am glad to believe has no original. Of the Indian of South America I know nothing; for I would not honor with the appelation of knowledge, what I derive from the fables published of them. These I believe to be just as true as the fables of Æsop. This belief is founded on what I have seen of man, white, red, and black, and what has been written of him by authors, enlightened themselves, and writing amidst an enlightened people. The Indian of North America being more within our reach, I can speak of him somewhat from my own knowledge, but more from the information of others better acquainted with him, and on whose truth and judgment I can rely. From these sources I am able to say, in contradiction to this representation, that he is neither more defective in ardor, nor more impotent with his female, than the white reduced to the same diet and exercise: that he is brave, when an enterprise depends on bravery; education with him making the point of honor consist in the destruction of an enemy by stratagem, and in the preservation of his own person free from injury; or perhaps this is nature; while it is education which teaches us to honor force more than finesse; that he will defend himself against an host of enemies, always chusing to be killed, rather than to surrender, though

[2]Quotation in French in the original. This translation follows that in *Notes on the State of Virginia*, ed. William Peden (Chapel Hill, 1955), 58–59.

it be to the whites, who he knows will treat him well: that in other situations also he meets death with more deliberation, and endures tortures with a firmness unknown almost to religious enthusiasm with us: that he is affectionate to his children, careful of them, and indulgent in the extreme: that his affections comprehend his other connections, weakening, as with us, from circle to circle, as they recede from the center: that his friendships are strong and faithful to the uttermost extremity: that his sensibility is keen, even the warriors weeping most bitterly on the loss of their children, though in general they endeavour to appear superior to human events: that his vivacity and activity of mind is equal to ours in the same situation; hence his eagerness for hunting, and for games of chance. The women are submitted to unjust drudgery. This I believe is the case with every barbarous people. With such, force is law. The stronger sex therefore imposes on the weaker. It is civilization alone which replaces women in the enjoyment of their natural equality. That first teaches us to subdue the selfish passions, and to respect those rights in others which we value in ourselves. Were we in equal barbarism, our females would be equal drudges. The man with them is less strong than with us, but their woman stronger than ours; and both for the same obvious reason; because our man and their woman is habituated to labour, and formed by it. With both races the sex which is indulged with ease is least athletic. An Indian man is small in the hand and wrist for the same reason for which a sailor is large and strong in the arms and shoulders, and a porter in the legs and thighs.—They raise fewer children than we do. The causes of this are to be found, not in a difference of nature, but of circumstance. The women very frequently attending the men in their parties of war and of hunting, child-bearing becomes extremely inconvenient to them. It is said, therefore, that they have learnt the practice of procuring abortion by the use of some vegetable; and that it even extends to prevent conception for a considerable time after. During these parties they are exposed to numerous hazards, to excessive exertions, to the greatest extremities of hunger. Even at their homes the nation depends for food, through a certain part of every year, on the gleanings of the forest: that is, they experience a famine once in every year. With all animals, if the female be badly fed, or not fed at all, her young perish: and if both male and female be reduced to like want, generation becomes less active, less productive. To the obstacles then of want and hazard, which nature has opposed to the multiplication of wild animals, for the purpose of restraining their numbers within certain bounds, those of labour and of voluntary abortion are added with the Indian. No wonder then if they multiply less than we do. Where food is regularly supplied, a single farm will shew more of cattle, than a whole country of forests can of buffaloes. The same Indian women, when married to white traders, who feed them and their children plentifully and regularly, who exempt them from excessive drudgery, who keep them stationary and unexposed to accident, produce and raise as many children as the white women. Instances are known, under these circumstances, of their rearing a dozen children. An inhuman practice once prevailed in this country of making slaves of the Indians. It is a fact well known with us, that the Indian women so enslaved produced and raised as numerous families as either the whites or blacks among whom they lived. — It has been said, that Indians have less hair than the whites, except on the head. But this is a fact of which fair proof can scarcely be had. With them it is disgraceful to be hairy on the body. They say it likens them to hogs. They therefore pluck the hair as fast as it appears. But the traders who marry their women, and prevail on them to discontinue this practice, say, that nature is the same with them as with the whites. Nor, if the fact be true, is the consequence necessary which has been drawn from it. Negroes have notoriously less hair than the whites; yet they are more ardent. But if cold and moisture be the agents of nature for diminishing the races of animals, how comes she all at once to suspend their operation as to the physical man of the new world, whom the Count [Buffon] acknowledges to be "about the same size as the man of our hemisphere,"[3] and to let loose their influence on his moral faculties? How has this "combination of the elements and other physical causes, so contrary to the enlargement of animal nature in this new world, these obstacles to the developement and formation of great germs," been arrested and suspended, so as to permit the human body to acquire its just dimensions, and by what inconceivable process has their action been directed on his mind alone? To judge of the truth of this, to form a just estimate of their genius and mental powers, more facts are wanting, and great allowance to be made for those circumstances of their situation which call for a display of particular talents only. This done, we shall probably find that they are formed in mind as well as in body, on the same module with the "Homo sapiens Europæus." . . .

Before we condemn the Indians of this continent as wanting genius, we must consider that letters have not yet been introduced among

[3]Quotation in French in the original

them. Were we to compare them in their present state with the Europeans North of the Alps, when the Roman arms and arts first crossed those mountains, the comparison would be unequal, because, at that time, those parts of Europe were swarming with numbers; because numbers produce emulation, and multiply the chances of improvement, and one improvement begets another. Yet I may safely ask, How many good poets, how many able mathematicians, how many great inventors in arts or sciences, had Europe North of the Alps then produced? And it was sixteen centuries after this before a Newton could be formed. I do not mean to deny, that there are varieties in the race of man, distinguished by their powers both of body and mind. I believe there are, as I see to be the case in the races of other animals. I only mean to suggest a doubt, whether the bulk and faculties of animals depend on the side of the Atlantic on which their food happens to grow, or which furnishes the elements of which they are compounded? Whether nature has enlisted herself as a Cis[4] or Trans-Atlantic partisan? I am induced to suspect, there has been more eloquence than sound reasoning displayed in support of this theory; that it is one of those cases where the judgment has been seduced by a glowing pen: and whilst I render every tribute of honor and esteem to the celebrated zoologist, who has added, and is still adding, so many precious things to the treasures of science, I must doubt whether in this instance he has not cherished error also, by lending her for a moment his vivid imagination and bewitching language.

So far the Count de Buffon has carried this new theory of the tendency of nature to belittle her productions on this side of the Atlantic. Its application to the race of whites, transplanted from Europe, remained for the Abbé Raynal. "One must be astonished (he says) that America has not yet produced one good poet, one able, mathematician, one man of genius in a single art or a single science."[5] "America has not yet produced one good poet." When we shall have existed as a people as long as the Greeks did before they produced a Homer, the Romans a Virgil, the French a Racine and Voltaire, the English a Shakespeare and Milton, should this reproach be still true, we will enquire from what unfriendly causes it has proceeded, that the other countries of Europe and quarters of the earth shall not have inscribed any name in the roll of poets. But neither has America produced "one able mathematician, one man of genius in a single art or a single

---

[4] *Cis:* on this side of the Atlantic
[5] Quotation in French in the original

science." In war we have produced a Washington, whose memory will be adored while liberty shall have votaries, whose name will triumph over time, and will in future ages assume its just station among the most celebrated worthies of the world, when that wretched philosophy shall be forgotten which would have arranged him among the degeneracies of nature. In physics we have produced a Franklin, than whom no one of the present age has made more important discoveries, nor has enriched philosophy with more, or more ingenious solutions of the phænomena of nature. We have supposed Mr. Rittenhouse second to no astronomer living: that in genius he must be the first, because he is self-taught. As an artist he has exhibited as great a proof of mechanical genius as the world has ever produced. He has not indeed made a world; but he has by imitation approached nearer its Maker than any man who has lived from the creation to this day. As in philosophy and war, so in government, in oratory, in painting, in the plastic art, we might shew that America, though but a child of yesterday, has already given hopeful proofs of genius, as well of the nobler kinds, which arouse the best feelings of man, which call him into action, which substantiate his freedom, and conduct him to happiness, as of the subordinate, which serve to amuse him only. We therefore suppose, that this reproach is as unjust as it is unkind; and that, of the geniuses which adorn the present age, America contributes its full share. . . .

QUERY XI: ABORIGINES

When the first effectual settlement of our colony was made, which was in 1607, the country from the sea-coast to the mountains, and from Patowmac [Potomac] to the most southern waters of James river, was occupied by upwards of forty different tribes of Indians. Of these the *Powhatans*, the *Mannahoacs*, and *Monacans*, were the most powerful. Those between the sea-coast and falls of the rivers, were in amity with one another, and attached to the *Powhatans* as their link of union. Those between the falls of the rivers and the mountains, were divided into two confederacies; the tribes inhabiting the head waters of Patowmac and Rappahanoc[k] being attached to the *Mannahoacs;* and those on the upper parts of James river to the *Monacans*. But the *Monacans* and their friends were in amity with the *Mannahoacs* and their friends, and waged joint and perpetual war against the *Powhatans*. We are told that the *Powhatans, Mannahoacs,* and *Monacans,* spoke languages so radically different, that interpreters were necessary when they transacted business. Hence we may conjecture, that this was not the case between all the tribes, and probably that each spoke the language of the nation to which it was attached; which we know to have been the case in many particular instances. Very possibly there may have been anciently three different stocks, each of which multiplying in a long course of time, had separated into so many little societies. This practice results from the circumstance of their having never submitted themselves to any laws, any coercive power, any shadow of government. Their only controuls are their manners, and that moral sense of right and wrong, which, like the sense of tasting and feeling, in every man makes a part of his nature. An offence against these is punished by contempt, by exclusion from society, or, where the case is serious, as that of murder, by the individuals whom it concerns. Imperfect as this species of coercion may seem, crimes are very rare among them: insomuch that were it made a question, whether no law, as among the savage Americans, or too much law, as among the civilized Europeans, submits man to the greatest evil, one who has seen both conditions of existence would pronounce it to be the last: and that the sheep are happier of themselves, than under care of the wolves. It will be said, that great societies cannot exist without government. The savages therefore break them into small ones. . . .

I know of no such thing existing as an Indian monument: for I would not honour with that name arrow points, stone hatchets, stone pipes, and half-shapen images. Of labour on the large scale, I think there is no remain as respectable as would be a common ditch for the draining of lands: unless indeed it be the Barrows,[6] which many are to be found all over this country. These are of different sizes, some of them constructed of earth, and some of loose stones. That they were repositories of the dead, has been obvious to all: but on what particular occasion constructed, was matter of doubt. Some have thought they covered the bones of those who have fallen in battles fought on the spot of interment. Some ascribed them to the custom, said to prevail among the Indians, of collecting, at certain periods, the bones of all their dead, wheresoever deposited at the time of death. Others again supposed them the general sepulchres for towns, conjectured to have been on or near these grounds; and this opinion was supported by the quality of the lands in which they are found, (those constructed of earth being generally in the softest and most fertile

[6]*Barrows:* burial mounds

meadow-grounds on river sides) and by a tradition, said to be handed down from the aboriginal Indians, that, when they settled in a town, the first person who died was placed erect, and earth put about him, so as to cover and support him; that, when another died, a narrow passage was dug to the first, the second reclined against him, and the cover of earth replaced, and so on. There being one of these in my neighbourhood, I wished to satisfy myself whether any, and which of these opinions were just. For this purpose I determined to open and examine it thoroughly. It was situated on the low grounds of the Rivanna, about two miles above its principal fork, and opposite to some hills, on which had been an Indian town. It was of a spheroidical form, of about 40 feet diameter at the base, and had been of about twelve feet altitude, though now reduced by the plough to seven and a half, having been under cultivation about a dozen years. Before this it was covered with trees of twelve inches diameter, and round the base was an excavation of five feet depth and width, from whence the earth had been taken of which the hillock was formed. I first dug superficially in several parts of it, and came to collections of human bones, at different depths, from six inches to three feet below the surface. These were lying in the utmost confusion, some vertical, some oblique, some horizontal, and directed to every point of the compass, entangled, and held together in clusters by the earth. Bones of the most distant parts were found together, as, for instance, the small bones of the foot in the hollow of a scull, many sculls would sometimes be in contact, lying on the face, on the side, on the back, top or bottom, so as, on the whole, to give the idea of bones emptied promiscuously from a bag or basket, and covered over with earth, without any attention to their order. The bones of which the greatest numbers remained, were sculls, jaw-bones, teeth, the bones of the arms, thighs, legs, feet, and hands. A few ribs remained, some vertebræ of the neck and spine, without their processes, and one instance only of the bone which serves as a base to the vertebral column. The sculls were so tender, that they generally fell to pieces on being touched. The other bones were stronger. There were some teeth which were judged to be smaller than those of an adult; a scull, which, on a slight view, appeared to be that of an infant, but it fell to pieces on being taken out, so as to prevent satisfactory examination; a rib, and a fragment of the under-jaw of a person about half grown; another rib of an infant; and part of the jaw of a child, which had not yet cut its teeth. This last furnishing the most decisive proof of the burial of children here, I was particular in my attention to it. It was part of the right-half of the under-jaw. The processes, by which it was articulated to the temporal bones, were entire; and the bone itself firm to where it had been broken off, which, as nearly as I could judge, was about the place of the eyetooth. Its upper edge, wherein would have been the sockets of the teeth, was perfectly smooth. Measuring it with that of an adult, by placing their hinder processes together, its broken end extended to the penultimate grinder of the adult. This bone was white, all the others of a sand colour. The bones of infants being soft, they probably decay sooner, which might be the cause so few were found here. I proceeded then to make a perpendicular cut through the body of the barrow, that I might examine its internal structure. This passed about three feet from its center, was opened to the former surface of the earth, and was wide enough for a man to walk through and examine its sides. At the bottom, that is, on the level of the circumjacent plain, I found bones; above these a few stones, brought from a cliff a quarter of a mile off, and from the river one-eighth of a mile off; then a large interval of earth, then a stratum of bones, and so on. At one end of the section were four strata of bones plainly distinguishable; at the other, three; the strata in one part not ranging with those in another. The bones nearest the surface were least decayed. No holes were discovered in any of them, as if made with bullets, arrows, or other weapons. I conjectured that in this barrow might have been a thousand skeletons. Every one will readily seize the circumstances above related, which militate against the opinion, that it covered the bones only of persons fallen in battle; and against the tradition also, which would make it the common sepulchre of a town, in which the bodies were placed upright, and touching each other. Appearances certainly indicate that it has derived both origin and growth from the accustomary[7] collection of bones, and deposition of them together; that the first collection had been deposited on the common surface of the earth, a few stones put over it, and then a covering of earth, that the second had been laid on this, had covered more or less of it in proportion to the number of bones, and was then also covered with earth; and so on. The following are the particular circumstances which give it this aspect. 1. The number of bones. 2. Their confused position. 3. Their being in different strata. 4. The strata in one part having no correspondence with those in another. 5. The different states of decay in these strata, which seem to indicate a difference in the time of inhumation. 6. The existence of infant bones among them.

[7]*accustomary:* customary, familiar

But on whatever occasion they may have been made, they are of considerable notoriety among the Indians: for a party passing, about thirty years ago, through the part of the country where this barrow is, went through the woods directly to it, without any instructions or enquiry, and having staid [stayed] about it some time, with expressions which were construed to be those of sorrow, they returned to the high road, which they had left about half a dozen miles to pay this visit, and pursued their journey. There is another barrow, much resembling this in the low grounds of the South branch of Shenandoah, where it is crossed by the road leading from the Rockfish gap to Staunton. Both of these have, within these dozen years, been cleared of their trees and put under cultivation, are much reduced in their height, and spread in width, by the plough, and will probably disappear in time. There is another on a hill in the Blue ridge of mountains, a few miles North of Wood's gap, which is made up of small stones thrown together. This has been opened and found to contain human bones, as the others do. There are also many others in other parts of the country. . . .

QUERY XIV: LAWS

. . . Many of the laws which were in force during the monarchy being relative merely to that form of government, or inculcating principles inconsistent with republicanism, the first assembly which met after the establishment of the commonwealth appointed a committee to revise the whole code, to reduce it into proper form and volume, and report it to the assembly. This work has been executed by three gentlemen, and reported; but probably will not be taken up till a restoration of peace shall leave to the legislature leisure to go through such a work. . . .

The following are the most remarkable alterations proposed:

To change the rules of descent, so as that the lands of any person dying intestate shall be divisible equally among all his children, or other representatives, in equal degree.

To make slaves distributable among the next of kin, as other moveables.

To have all public expences, whether of the general treasury, or of a parish or county, (as for the maintenance of the poor, building bridges, court-houses, &c.) supplied by assessments on the citizens, in proportion to their property.

To hire undertakers for keeping the public roads in repair, and indemnify individuals through whose lands new roads shall be opened.

To define with precision the rules whereby aliens should become citizens, and citizens make themselves aliens.

To establish religious freedom on the broadest bottom.[8]

To emancipate all slaves born after passing the act. The bill reported by the revisors does not itself contain this proposition; but an amendment containing it was prepared, to be offered to the legislature whenever the bill should be taken up, and further directing, that they should continue with their parents to a certain age, then be brought up, at the public expence, to tillage, arts or sciences, according to their geniusses, till the females should be eighteen, and the males twenty-one years of age, when they should be colonized to such place as the circumstances of the time should render most proper, sending them out with arms, implements of houshold and of the handicraft arts, seeds, pairs of the useful domestic animals, &c. to declare them a free and independent people, and extend to them our alliance and protection, till they shall have acquired strength; and to send vessels at the same time to other parts of the world for an equal number of white inhabitants; to induce whom to migrate hither, proper encouragements were to be proposed. It will probably be asked, Why not retain and incorporate the blacks into the state, and thus save the expence of supplying, by importation of white settlers, the vacancies they will leave? Deep rooted prejudices entertained by the whites; ten thousand recollections, by the blacks, of the injuries they have sustained; new provocations; the real distinctions which nature has made; and many other circumstances, will divide us into parties, and produce convulsions which will probably never end but in the extermination of the one or the other race.—To these objections, which are political, may be added others, which are physical and moral. The first difference which strikes us is that of colour. Whether the black of the negro resides in the reticular membrane between the skin and scarf-skin, or in the scarf-skin itself; whether it proceeds from the colour of the blood, the colour of the bile, or from that of some other secretion, the difference is fixed in nature, and is as real as if its seat and cause were better known to us. And is this difference of no importance? Is it not the foundation of a greater or less share of beauty in the two races? Are not the fine mixtures of red and white, the expressions of every passion by greater or less suffusions of colour in the one, preferable

[8]*bottom:* basis

to that eternal monotony, which reigns in the countenances, that immoveable veil of black which covers all the emotions of the other race? Add to these, flowing hair, a more elegant symmetry of form, their own judgment in favour of the whites, declared by their preference of them, as uniformly as is the preference of the Oranootan [orangutan] for the black women over those of his own species. The circumstance of superior beauty, is thought worthy attention in the propagation of our horses, dogs, and other domestic animals; why not in that of man? Besides those of colour, figure, and hair, there are other physical distinctions proving a difference of race. They have less hair on the face and body. They secrete less by the kidnies, and more by the glands of the skin, which gives them a very strong and disagreeable odour. This greater degree of transpiration renders them more tolerant of heat, and less so of cold, than the whites. Perhaps too a difference of structure in the pulmonary apparatus, which a late ingenious experimentalist has discovered to be the principal regulator of animal heat, may have disabled them from extricating, in the act of inspiration, so much of that fluid from the outer air, or obliged them in expiration, to part with more of it. They seem to require less sleep. A black, after hard labour through the day, will be induced by the slightest amusements to sit up till midnight, or later, though knowing he must be out with the first dawn of the morning. They are at least as brave, and more adventuresome. But this may perhaps proceed from a want of fore-thought, which prevents their seeing a danger till it be present. When present, they do not go through it with more coolness or steadiness than the whites. They are more ardent after their female: but love seems with them to be more an eager desire, than a tender delicate mixture of sentiment and sensation. Their griefs are transient. Those numberless afflictions, which render it doubtful whether heaven has given life to us in mercy or in wrath, are less felt, and sooner forgotten with them. In general, their existence appears to participate more of sensation than reflection. To this must be ascribed their disposition to sleep when abstracted from their diversions, and unemployed in labour. An animal whose body is at rest, and who does not reflect, must be disposed to sleep of course. Comparing them by their faculties of memory, reason, and imagination, it appears to me, that in memory they are equal to the whites; in reason much inferior, as I think one could scarcely be found capable of tracing and comprehending the investigations of Euclid; and that in imagination they are dull, tasteless, and anomalous. It would be unfair to follow them to Africa for this investigation. We will consider them here, on the same stage with the whites, and where the facts are not apocryphal on which a judgment is to be formed. It will be right to make great allowances for the difference of condition, of education, of conversation, of the sphere in which they move. Many millions of them have been brought to, and born in America. Most of them indeed have been confined to tillage, to their own homes, and their own society: yet many have been so situated, that they might have availed themselves of the conversation of their masters; many have been brought up to the handicraft arts, and from that circumstance have always been associated with the whites. Some have been liberally educated, and all have lived in countries where the arts and sciences are cultivated to a considerable degree, and have had before their eyes samples of the best works from abroad. The Indians, with no advantages of this kind, will often carve figures on their pipes not destitute of design and merit. They will crayon out an animal, a plant, or a country, so as to prove the existence of a germ in their minds which only wants cultivation. They astonish you with strokes of the most sublime oratory; such as prove their reason and sentiment strong, their imagination glowing and elevated. But never yet could I find that a black had uttered a thought above the level of plain narration; never see even an elementary trait of painting or sculpture. In music they are more generally gifted than the whites with accurate ears for tune and time, and they have been found capable of imagining a small catch.[9] Whether they will be equal to the composition of a more extensive run of melody, or of complicated harmony, is yet to be proved. Misery is often the parent of the most affecting touches in poetry.—Among the blacks is misery enough, God knows, but no poetry. . . .

The opinion, that they are inferior in the faculties of reason and imagination, must be hazarded with great diffidence. To justify a general conclusion, requires many observations, even where the subject may be submitted to the anatomical knife, to optical glasses, to analysis by fire, or by solvents. How much more then where it is a faculty, not a substance, we are examining; where it eludes the research of all the senses; where the conditions of its existence are various and variously combined; where the effects of those which are present or absent bid defiance to calculation; let me add too, as a circumstance of great tenderness, where our conclusion would degrade a whole race of men from the rank in the scale of beings which their Creator may perhaps have given them. To our reproach it must be said, that

---

[9]*catch:* a round for three or more unaccompanied voices, with one continuous melody

though for a century and a half we have had under our eyes the races of black and of red men, they have never yet been viewed by us as subjects of natural history. I advance it therefore as a suspicion only, that the blacks, whether originally a distinct race, or made distinct by time and circumstances, are inferior to the whites in the endowments both of body and mind. It is not against experience to suppose, that different species of the same genus, or varieties of the same species, may possess different qualifications. Will not a lover of natural history then, one who views the gradations in all the races of animals with the eye of philosophy, excuse an effort to keep those in the department of man as distinct as nature has formed them? This unfortunate difference of colour, and perhaps of faculty, is a powerful obstacle to the emancipation of these people. Many of their advocates, while they wish to vindicate the liberty of human nature, are anxious also to preserve its dignity and beauty. Some of these, embarrassed by the question "What further is to be done with them?" join themselves in opposition with those who are actuated by sordid avarice only. Among the Romans emancipation required but one effort. The slave, when made free, might mix with, without staining the blood of his master. But with us a second is necessary, unknown to history. When freed, he is to be removed beyond the reach of mixture. . . .

24

TIMOTHY DWIGHT

Greenfield Hill: A Poem in Seven Parts
1794

Timothy Dwight (1752–1817) was a Yale graduate, prominent Congregational minister, and author of America's first epic poem, The Conquest of Canäan *(see Document 2). His extended pastoral poem* Greenfield Hill *was published in New York in 1794. It was Dwight's most popular poem, well received by critics and readers alike. It was also his last. In 1795, Dwight became president of Yale College, a position he held until his death.*

Timothy Dwight, *Greenfield Hill: A Poem in Seven Parts* (New York, 1794), part 4, 93–105, lines 188–81, 235–52, 316–60.

PART IV: THE DESTRUCTION OF THE PEQUODS[10]

. . .

Oft have I heard the tale, when matron sere
Sung to my infant ear the song of woe;
Of maiden meek, consum'd with pining care,
Around whose tomb the wild-rose lov'd to blow;
Or told, with swimming eyes, how, long ago,
Remorseless Indians, all in midnight dire,
The little, sleeping village, did o'erthrow,
Bidding the cruel flames to heaven aspire,
And scalp'd the hoary head, and burn'd the babe with fire.

Then, fancy-fir'd, her memory wing'd it's flight,
To long-forgotten wars, and dread alarms,
To chiefs obscure, but terrible in fight,
Who mock'd each foe, and laugh'd at deadliest harms,
Sydneys[11] in zeal, and Washingtons in arms.
By instinct tender to the woes of man,
My heart bewildering with sweet pity's charms,
Thro' solemn scenes, with Nature's step, she ran,
And hush'd her audience small, and thus the tale began.

"Thro' verdant banks where Thames's[12] branches glide,
Long held the Pequods an extensive sway;
Bold, savage, fierce, of arms the glorious pride,
And bidding all the circling realms obey.
Jealous, they saw the tribes, beyond the sea,
Plant in their climes; and towns, and cities, rise;
Ascending castles foreign flags display;
Mysterious art new scenes of life devise;
And steeds insult the plains, and cannon rend the skies."

"They saw, and soon the strangers' fate decreed,
And soon of war disclos'd the crimson sign;
First, hapless Stone! they bade thy bosom bleed,
A guiltless offering, at th' infernal shrine:
Then, gallant Norton! the hard fate was thine,

[10]Pequots
[11]Algernon Sydney (1622–1683), English writer and staunch opponent of monarchical power
[12]*Thames:* a river in southeastern Connecticut

By ruffians butcher'd, and denied a grave:
Thee, generous Oldham! next the doom malign
Arrested; nor could all thy courage save;
Forsaken, plunder'd, cleft, and buried in the wave."

"Soon the sad tidings reach'd the general ear;
And prudence, pity, vengeance, all inspire:
Invasive war their gallant friends prepare;
And soon a noble band, with purpose dire,
And threatening arms, the murderous fiends require:
Small was the band, but never taught to yield;
Breasts fac'd with steel, and souls instinct with fire:
Such souls, from Sparta, Persia's world repell'd,
When nations pav'd the ground, and Xerxes flew the field."[13]

"The rising clouds the Savage Chief descried,
And, round the forest, bade his heroes arm;
To arms the painted warriors proudly hied,
And through surrounding nations rung the alarm.
The nations heard; but smil'd, to see the storm,
With ruin fraught, o'er Pequod mountains driven;
And felt infernal joy the bosom warm,
To see their light hang o'er the skirts of even,
And other suns arise, to gild a kinder heaven."

"Swift to the Pequod fortress Mason[14] sped,
Far in the wildering wood's impervious gloom;
A lonely castle, brown with twilight dread;
Where oft th' embowel'd captive met his doom,
And frequent heav'd, around the hollow tomb;
Scalps hung in rows, and whitening bones were strew'd;
Where, round the broiling babe, fresh from the womb,
With howls the Powaw[15] fill'd the dark abode,
And screams, and midnight prayers, invok'd the Evil god."

. . .

"On the drear walls a sudden splendour glow'd,
There Mason shone, and there his veterans pour'd.
Anew the Hero claim'd the fiends of blood,

---

[13]Xerxes I (519?–465 B.C.), King of Persia, attacked Greece with a huge force that was decisively defeated by an alliance of 30 Greek states led by Sparta.
[14]John Mason of Connecticut led the English and Mohegan attack on the Pequots.
[15]powwow

While answering storms of arrows round him shower'd,
And the war-scream the ear with anguish gor'd.
Alone, he burst the gate: the forest round
Re-echoed death; the peal of onset roar'd;
In rush'd the squadrons; earth in blood was drown'd;
And gloomy spirits fled, and corses hid the ground."

"Not long in dubious fight the host had striven,
When, kindled by the musket's potent flame,
In clouds, and fire, the castle rose to heaven,
And gloom'd the world, with melancholy beam.
Then hoarser groans, with deeper anguish, came;
And fiercer fight the keen assault repell'd:
Nor even these ills the savage breast could tame;
Like hell's deep caves, the hideous region yell'd,
'Till death, and sweeping fire, laid waste the hostile field."

. . .

Fierce, dark, and jealous, is the exotic soul,
That, cell'd in secret, rules the savage breast.
There treacherous thoughts of gloomy vengeance roll,
And deadly deeds of malice unconfess'd;
The viper's poison rankling in it's nest.
Behind his tree, each Indian aims unseen:
No sweet oblivion soothes the hate impress'd:
Years fleet in vain: in vain realms intervene:
The victim's blood alone can quench the flames within.

Their knives the tawny tribes in slaughter steep,
When men, mistrustless, think them distant far;
And, when blank midnight shrouds the world in sleep,
The murderous yell announces first the war.
In vain sweet smiles compel the fiends to spare;
Th' unpitied victim screams, in tortures dire;
The life-blood stains the virgin's bosom bare;
Cherubic infants, limb by limb expire;
And silver'd Age sinks down in slowly-curling fire.

Yet savages are men. With glowing heat,
Fix'd as their hatred, friendship fills their mind;
By acts with justice, and with truth, replete,
Their iron breasts to softness are inclin'd.
But when could War of converts boast refin'd?

Or when Revenge to peace and sweetness move?
His heart, man yields alone to actions kind;
His faith, to creeds, whose soundness virtues prove,
Thawn in the April sun, and opening still to love.

Senate august! that sway'st Columbian climes,
Form'd of the wise, the noble, and humane,
Cast back the glance through long-ascending times,
And think what nations fill'd the western plain.
Where are they now? What thoughts the bosom pain,
From mild Religion's eye how streams the tear,
To see so far outspread the waste of man,
And ask "How fell the myriads, HEAVEN plac'd here!"
Reflect, be just, and feel for Indian woes severe.

But cease, foul Calumny! with sooty tongue,
No more the glory of our sires belie.
They felt, and they redress'd, each nation's wrong;
Even Pequod foes they view'd with generous eye,
And, pierc'd with injuries keen, that Virtue try,
The savage faith, and friendship, strove to gain:
And, had no base Canadian fiends been nigh,
Even now soft Peace had smil'd on every plain,
And tawny nations liv'd, and own'd MESSIAH's reign. . . .

JOHN VANDERLYN

The Death of Jane McCrea
1804

A native of New York, John Vanderlyn (1775–1852) worked for a while as the painter Gilbert Stuart's assistant in Philadelphia. In 1796, he traveled to Paris, where he studied at the École des Beaux-Arts and completed a series of history paintings. The first of these, *The Death of Jane McCrea* of 1804, is based on an event that received widespread attention during and after the Revolution. Jane McCrea, an American woman traveling through enemy lines to meet the English army officer she intended to marry, was murdered by two Mohawk warriors in July 1777. Her death was remembered for many decades in popular ballads, poems, and periodicals in America.
Wadsworth Atheneum, Hartford. Purchased by Wadsworth Museum.

WILLIAM BARTRAM

Travels through North & South Carolina, Georgia, East & West Florida, the Cherokee Country, the Extensive Territories of the Muscogulges[16] or Creek Confederacy, and the Country of the Chactaws[17]

1791

William Bartram (1739–1823), son of the Philadelphia botanist John Bartram, was the most respected American naturalist of his generation. He entertained a steady stream of prominent visitors to his father's botanical garden, was elected to membership in the American Philosophical Society of Philadelphia, and was invited by President Thomas Jefferson to serve as naturalist on a government-sponsored expedition up the Red River in 1803. (Bartram declined because of advancing age and ill health.) His Travels, *published in Philadelphia in 1791, record his four-year journey of 20 years before. Reprinted within the decade in England, Germany, Ireland, Holland, and France, the book reached an international audience and influenced naturalists throughout Europe and America.*

PART I.

Chap. III.

. . . I arrived at St. Ille's in the evening, where I lodged, and next morning having crossed over in a ferry boat, sat forward for St. Mary's. The situation of the territory, it's soil and productions, between these two last rivers, are nearly similar to those which I had passed over, except that the savannas are more frequent and extensive.

[16]Muskogees
[17]Choctaws

William Bartram, *Travels through North & South Carolina, Georgia, East & West Florida, the Cherokee Country, the Extensive Territories of the Muscogulges or Creek Confederacy, and the Country of the Chactaws* (Philadelphia, 1791), frontispiece, 20–23, 483–93.

Frontispiece to Bartram's *Travels:* Mico Chlucco the Long Warrior, or King of the Siminoles.

It may be proper to observe, that I had now passed the utmost frontier of the white settlements on that border. It was drawing on towards the close of day, the skies serene and calm, the air temperately cool, and gentle zephyrs breathing through the fragrant pines; the prospect around enchantingly varied and beautiful; endless green savannas, checquered with coppices of fragrant shrubs, filled the air with the richest perfume. The gaily attired plants which enamelled the green had begun to imbibe the pearly dew of evening; nature seemed silent, and nothing appeared to ruffle the happy moments of evening contemplation; when, on a sudden, an Indian appeared crossing the path, at a considerable distance before me. On perceiving that he was armed with a rifle, the first sight of him startled me, and I endeavoured to elude his sight, by stopping my pace, and keeping large trees between us; but he espied me, and turning short about, sat spurs to his horse, and came up on full gallop. I never before this was afraid at the sight of an Indian, but at this time, I must own that my spirits were very much agitated: I saw at once, that being unarmed, I was in his power, and having now but a few moments to prepare, I resigned myself entirely to the will of the Almighty, trusting to his mercies for my preservation; my mind then became tranquil, and I resolved to meet the dreaded foe with resolution and chearful confidence. The intrepid Siminole stopped suddenly, three or four yards before me, and silently viewed me, his countenance angry and fierce, shifting his rifle from shoulder to shoulder, and looking about instantly on all sides. I advanced towards him, and with an air of confidence offered him my hand, hailing him, brother; at this he hastily jerked back his arm, with a look of malice, rage and disdain, seeming every way disconcerted; when again looking at me more attentively, he instantly spurred up to me, and, with dignity in his look and action, gave me his hand. Possibly the silent language of his soul, during the moment of suspense (for I believe his design was to kill me when he first came up) was after this manner: "White man, thou art my enemy, and thou and thy brethren may have killed mine; yet it may not be so, and even were that the case, thou art now alone, and in my power. Live; the Great Spirit forbids me to touch thy life; go to thy brethren, tell them thou sawest an Indian in the forests, who knew how to be humane and compassionate." In fine, we shook hands, and parted in a friendly manner, in the midst of a dreary wilderness; and he informed me of the course and distance to the trading-house, where I found he had been extremely ill treated the day before.

I now sat forward again, and after eight or ten miles riding, arrived at the banks of St. Mary's, opposite the stores, and got safe over before dark. The river is here about one hundred yards across, has ten feet water, and, following its course, about sixty miles to the sea, though but about twenty miles by land. The trading company here received and treated me with great civility. On relating my adventures on the road, particularly the last with the Indian, the chief replied, with a countenance that at once bespoke surprise and pleasure, "My friend, consider yourself a fortunate man: that fellow," said he, "is one of the greatest villains on earth, a noted murderer, and outlawed by his countrymen. Last evening he was here, we took his gun from him, broke it in pieces, and gave him a severe drubbing: he, however, made his escape, carrying off a new rifle gun, with which, he said, going off, he would kill the first white man he met."

On seriously contemplating the behaviour of this Indian towards me, so soon after his ill treatment, the following train of sentiments insensibly crouded in upon my mind.

Can it be denied, but that the moral principle, which directs the savages to virtuous and praiseworthy actions, is natural or innate? It is certain they have not the assistance of letters, or those means of education in the schools of philosophy, where the virtuous sentiments and actions of the most illustrious characters are recorded, and carefully laid before the youth of civilized nations: therefore this moral principle must be innate, or they must be under the immediate influence and guidance of a more divine and powerful preceptor, who, on these occasions, instantly inspires them, and as with a ray of divine light, points out to them at once the dignity, propriety, and beauty of virtue.

The land on, and adjacent to, this river, notwithstanding its arenaceous[18] surface, appears naturally fertile. The Peach trees are large, healthy, and fruitful; and Indian Corn, Rice, Cotton, and Indigo, thrive exceedingly. This sandy surface, one would suppose, from it's loose texture, would possess a percolating quality, and suffer the rainwaters quickly to drain off; but it is quite the contrary, at least in these low maritime sandy countries of Carolina and Florida, beneath the mountains; for in the sands, even the heights, where the arenaceous stratum is perhaps five, eight, and ten feet above the clay, the earth, even in the longest droughts, is moist an inch or two under the surface; whereas, in the rich tenacious low lands, at such times, the ground is dry, and, as it were, baked many inches, and sometimes some feet deep, and the crops, as well as almost all vegetation, suffer in such soils and situations. The reason of this may be, that this kind of earth admits more freely of a transpiration of vapours, arising from intestine[19] watery canals to the surface; and probably these vapours are impregnated with saline or nitrous principles, friendly and nutritive

---

[18]*arenaceous:* sandy
[19]*intestine:* underground

to vegetables; however, of these causes and secret operations of nature I am ignorant, and resume again my proper employment, that of discovering and collecting data for the exercise of more able physiologists. . . .

PART IV.

An account of the persons, manners, customs and government, of the Muscogulges or Creeks, Cherokees, Chactaws, &c. Aborigines of the Continent of North America

Chap. I.

Description of the character, customs and persons of the American aborigines, from my own observations, as well as from the general and impartial report of ancient, respectable men, either of their own people, or white traders, who have spent many days of their lives amongst them.

PERSONS AND QUALIFICATIONS

The males of the Cherokees, Muscogulges, Siminoles, Chic[k]asaws, Chactaws [Choctaws], and confederate tribes of the Creeks, are tall, erect, and moderately robust; their limbs well shaped, so as generally to form a perfect human figure; their features regular, and countenance open, dignified and placid; yet the forehead and brow so formed, as to strike you instantly with heroism and bravery; the eye though rather small, yet active and full of fire; the iris always black, and the nose commonly inclining to the aquiline.

Their countenance and actions exhibit an air of magnanimity, superiority and independence.

Their complexion, of a reddish brown or copper colour; their hair long, lank, coarse, and black as a raven, and reflecting the like lustre at different exposures to the light.

The women of the Cherokees, are tall, slender, erect and of a delicate frame, their features formed with perfect symmetry, their countenance cheerful and friendly, and they move with a becoming grace and dignity.

The Muscogulge women, though remarkably short of stature, are well formed; their visage round, features regular and beautiful; the brow high and arched; the eye large, black and languishing, expressive of modesty, diffidence, and bashfulness; these charms are their defensive and offensive weapons, and they know very well how to play them off, and under cover of these alluring graces, are concealed the most subtile artifice; they are however loving and affectionate. They are, I believe, the smallest race of women yet known, seldom above five feet high, and I believe the greater number never arrive to that stature; their hands and feet not larger than those of Europeans of nine or ten years of age. Yet the men are of gigantic stature, a full size larger than Europeans; many of them above six feet, and few under that, or five feet eight or ten inches. Their complexion much darker than any of the tribes to the North of them that I have seen. This description will I believe comprehend the Muscogulges, their confederates, the Chactaws, and I believe the Chicasaws (though I have never seen their women), excepting however some bands of the Siminoles, Uches [Uchees] and Savannucas [Savannas], who are rather taller and slenderer, and their complexion brighter.

The Cherokees are yet taller and more robust than the Muscogulges, and by far the largest race of men I have seen;* their complexions brighter and somewhat of the olive cast, especially the adults; and some of their young women are nearly as fair and blooming as European women.

The Cherokees in their dispositions and manners are grave and steady; dignified and circumspect in their deportment; rather slow and reserved in conversation; yet frank, cheerful, and humane; tenacious of the liberties and natural rights of man; secret, deliberate and determined in their councils; honest, just and liberal, and ready always to sacrifice every pleasure and gratification, even their blood, and life itself, to defend their territory and maintain their rights. They do homage to the Muscogulges with reluctance, and are impatient under that galling yoke. I was witness to a most humiliating lash, which they passively received from their red masters, at the great congress and treaty of Augusta, when these people acceded with the Creeks, to the cession of the New Purchase;[20] where were about three hundred of the Creeks, a great part of whom were warriors, and about one hundred Cherokees.

The first day of convention opened with settling the preliminaries, one article of which was a demand on the part of the Georgians, to a

*There are, however, some exceptions to this general observation, as I have myself witnessed. Their present grand chief or emperor (the Little Carpenter, Atta-kul-kulla) is a man of remarkably small stature, slender, and of a delicate frame, the only instance I saw in the nation: but he is a man of superior abilities.
[20]*New Purchase:* Territory ceded by the Indians at Augusta, Georgia, in 1773

territory lying on the Tugilo [Tugaloo River], and claimed by them both, which it seems the Cherokees had, previous to the opening of congress, privately conveyed to the Georgians, unknown to the Creeks. The Georgians mentioning this as a matter settled, the Creeks demanded in council, on what foundation they built that claim, saying they had never ceded these lands. The Georgians answered, that they bought them of their friends and brothers the Cherokees. The Creeks nettled and incensed at this, a chief and warrior started up, and with an agitated and terrific countenance, frowning menaces and disdain, fixed his eyes on the Cherokee chiefs, and asked them what right they had to give away their lands, calling them old women, and saying they had long ago obliged them to wear the petticoat; a most humiliating and degrading stroke, in the presence of the chiefs of the whole Muscogulge confederacy, of the Chicasaws, principal men and citizens of Georgia, Carolina, Virginia, Maryland and Pennsylvania, in the face of their own chiefs and citizens, and amidst the laugh and jeers of the assembly, especially the young men of Virginia, their old enemy and dreaded neighbour: but humiliating as it really was, they were obliged to bear the stigma passively, and even without a reply.

And moreover, these arrogant bravos and usurpers carried their pride and importance to such lengths, as even to threaten to dissolve the congress and return home, unless the Georgians consented to annul the secret treaty with the Cherokees, and receive that territory immediately from them, as acknowledging their exclusive right of alienation; which was complied with, though violently extorted from the Cherokees, contrary to right and sanction of treaties; since the Savanna[h] river and its waters were acknowledged to be the natural and just bounds of territory betwixt the Cherokees and Muscogulges.

The national character of the Muscogulges, when considered in a political view, exhibits a portraiture of a great or illustrious heroe. A proud, haughty and arrogant race of men; they are brave and valiant in war, ambitious of conquest, restless and perpetually exercising their arms, yet magnanimous and merciful to a vanquished enemy, when he submits and seeks their friendship and protection: always uniting the vanquished tribes in confederacy with them; when they immediately enjoy, unexceptionably, every right of free citizens, and are from that moment united in one common band of brotherhood. They were never known to exterminate a tribe, except the Yamasees, who would never submit on any terms, but fought it out to the last, only about forty or fifty of them escaping at the last decisive battle, who threw themselves under the protection of the Spaniards at St. Augustine.

According to their own account, which I believe to be true, after their arrival in this country, they joined in alliance and perpetual amity with the British colonists of South Carolina and Georgia, which they never openly violated; but on the contrary, pursued every step to strengthen the alliance; and their aged chiefs to this day, speak of it with tears of joy, and exult in that memorable transaction, as one of the most glorious events in the annals of their nation.

As an instance of their ideas of political impartial justice, and homage to the Supreme Being, as the high arbiter of human transactions, who alone claims the right of taking away the life of man, I beg leave to offer to the reader's consideration, the following event, as I had it from the mouth of a Spaniard, a respectable inhabitant of East Florida.

The son of the Spanish governor of St. Augustine, together with two young gentlemen, his friends and associates, conceived a design of amusing themselves in a party of sport, at hunting and fishing. Having provided themselves with a convenient bark, ammunition, fishing tackle, &c., they set sail, directing their course South, along the coast, towards the point of Florida, putting into bays and rivers, as conveniency and the prospect of game invited them. The pleasing rural and diversified scenes of the Florida coast, imperceptibly allured them far to the south, beyond the Spanish fortified post. Unfortunate youths! regardless of the advice and injunctions of their parents and friends, still pursuing the delusive objects, they entered a harbour at evening, with a view of chasing the roe-buck, and hunting up the sturdy bear, solacing themselves with delicious fruits, and reposing under aromatic shades; when, alas! cruel unexpected event! in the beatific moments of their slumbers, they were surrounded, arrested and carried off by a predatory band of Creek Indians, proud of the capture, so rich a prize; they hurry away into cruel bondage the hapless youths, conducting them by devious paths through dreary swamps and boundless savannas, to the Nation.

At that time the Indians were at furious war with the Spaniards, scarcely any bounds set to their cruelties on either side: in short, the miserable youths were condemned to be burnt.

But there were English traders in these towns, who learning the character of the captives, and expecting great rewards from the Spanish governor, if they could deliver them, petitioned the Indians on their behalf, expressing their wishes to obtain their rescue, offering a great ransom; acquainting them at the same time, that they were young men of high rank, and one of them the governor's son.

Upon this, the head men, or chiefs of the whole nation, were convened, and after solemn and mature deliberation, they returned the traders their final answer and determination, which was as follows:

"Brothers and friends. We have been considering upon this business concerning the captives — and that, under the eye and fear of the Great Spirit. You know that these people are our cruel enemies; they save no lives of us red men, who fall in their power. You say that the youth is the son of the Spanish governor; we believe it; we are sorry he has fallen into our hands, but he is our enemy; the two young men (his friends) are equally our enemies; we are sorry to see them here; but we know no difference in their flesh and blood; they are equally our enemies; if we save one we must save all three: but we cannot do it; the red men require their blood to appease the spirits of their slain relatives; they have entrusted us with the guardianship of our laws and rights, we cannot betray them.

"However, we have a sacred prescription relative to this affair, which allows us to extend mercy to a certain degree: a third is saved by lot; the Great Spirit allows us to put it to that decision; he is no respecter of persons." The lots were cast. The governor's son was taken and burnt.

If we consider them with respect to their private character or in a moral view, they must, I think, claim our approbation, if we divest ourselves of prejudice and think freely. As moral men they certainly stand in no need of European civilization.

They are just, honest, liberal and hospitable to strangers; considerate, loving and affectionate to their wives and relations; fond of their children; industrious, frugal, temperate and persevering; charitable and forbearing. I have been weeks and months amongst them and in their towns, and never observed the least sign of contention or wrangling: never saw an instance of an Indian beating his wife, or even reproving her in anger. In this case they stand as examples of reproof to the most civilized nations, as not being defective in justice, gratitude and a good understanding; for indeed their wives merit their esteem and the most gentle treatment, they being industrious, frugal, careful, loving and affectionate.

The Muscogulges are more volatile, sprightly and talkative than their Northern neighbours, the Cherokees; and, though far more distant from the white settlements than any nation East of the Mississipi or Ohio, appear evidently to have made greater advances towards the refinements of true civilization, which cannot, in the least degree, be attributed to the good examples of the white people.

Their internal police and family economy at once engage the notice of European travellers, and incontrovertibly place these people in an illustrious point of view: their liberality, intimacy and friendly intercourse one with another, without any restraint of ceremonious formality, as if they were even insensible of the use or necessity of associating the passions or affections of avarice, ambition or covetousness.

A man goes forth on his business or avocations; he calls in at another town; if he wants victuals, rest or social conversation, he confidently approaches the door of the first house he chooses, saying "I am come;" the good man or woman replies, "You are; its well." Immediately victuals and drink are ready; he eats and drinks a little, then smokes tobacco, and converses either of private matters, public talks, or the news of the town. He rises and says, "I go!" the other answers, "You do!" He then proceeds again, and steps in at the next habitation he likes, or repairs to the public square, where are people always conversing by day, or dancing all night, or to some more private assembly, as he likes; he needs no one to introduce him, any more than the black-bird or thrush, when he repairs to the fruitful groves, to regale on their luxuries, and entertain the fond female with evening songs.

It is astonishing, though a fact, as well as a sharp reproof to the white people, if they will allow themselves liberty to reflect and form a just estimate, and I must own elevates these people to the first rank amongst mankind, that they have been able to resist the continual efforts of the complicated host of vices, that have for ages over-run the nations of the old world, and so contaminated their morals; yet more so, since such vast armies of these evil spirits have invaded this continent, and closely invested them on all sides. Astonishing indeed! when we behold the ill, immoral conduct of too many white people, who reside amongst them: notwithstanding which, it seems natural, eligible, and even easy, for these simple, illiterate people, to put in practice those beautiful lectures delivered to us by the ancient sages and philosophers, and recorded for our instruction.

I saw a young Indian in the Nation, who when present, and beholding the scenes of mad intemperance and folly acted by the white men in the town, clap his hand to his breast, and with a smile, looked aloft as if struck with astonishment, and wrapt in love and adoration to the Deity; as who should say, "O thou Great and Good Spirit! we are indeed sensible of thy benignity and favour to us red men, in denying us the understanding of white men. We did not know before they came amongst us that mankind could become so base, and fall so below the dignity of their nature. Defend us from their manners, laws and power."

The Muscogulges, with their confederates, the Chactaws, Chicasaws, and perhaps the Cherokees, eminently deserve the encomium of all nations, for their wisdom and virtue in resisting and even repelling the greatest, and even the common enemy of mankind, at least of most of the European nations, I mean spirituous liquors.

The first and most cogent article in all their treaties with the white people, is, that there shall not be any kind of spirituous liquors sold or brought into their towns; and the traders are allowed but two kegs (five gallons each) which is supposed to be sufficient for a company, to serve them on the road; and if any of this remains on their approaching the towns, they must spill it on the ground or secrete it on the road, for it must not come into the town.

On my journey from Mobile to the Nation, just after we had passed the junction of the Pensacola road with our path, two young traders overtook us on their way to the Nation. We inquired what news? They informed us that they were running about forty kegs of Jamaica spirits (which by dashing would have made at least eighty kegs) to the Nation; and after having left the town three or four days, they were surprised on the road in the evening, just after they had come to camp, by a party of Creeks, who discovering their species of merchandize, they forthwith struck their tomahawks into every keg, giving the liquor to the thirsty sand, not tasting a drop of it themselves; and they had enough to do to keep the tomahawks from their own skulls.

How are we to account for their excellent policy in civil government; it cannot derive its influence from coercive laws, for they have no such artificial system. Divine wisdom dictates and they obey.

We see and know full well the direful effects of this torrent of evil, which has its source in hell; and we know surely, as well as these savages, how to divert its course and suppress its inundations. Do we want wisdom and virtue? let our youth then repair to the venerable councils of the Muscogulges.

A Selected Cultural Chronology of the United States of America (1771–1806)

1771   Philip Freneau and Hugh Henry Brackenridge deliver "The Rising Glory of America" at the College of New Jersey's commencement in September.

1775   The Continental Congress passes a resolution against plays and "other expensive diversions and entertainments."

1776   Abigail Adams asks her husband, John, to "Remember the Ladies" in America's new code of laws in March; John Adams responds in April with a seriocomic warning about social revolution.

1779   The Supreme Executive Council of Pennsylvania commissions Charles Willson Peale to paint *George Washington at the Battle of Princeton*.

1783   Noah Webster publishes his speller, the first American textbook, as part of his *Grammatical Institute of the English Language*.

1785   Timothy Dwight publishes *The Conquest of Canäan; A Poem, in Eleven Books*. The Common Council of New York City closes the John Street Theater as a "fruitful source of dissipation, immorality, and vice."

1786   Noah Webster and Benjamin Franklin correspond about orthographic reform and the creation of an American language. Benjamin Rush calls for the creation of "republican machines" in his *Plan for the Establishment of Public Schools and the Diffusion of Knowledge in Pennsylvania; to Which are Added, Thoughts Upon the Mode of Education Proper in a Republic*. From Paris, Thomas Jefferson urges passage of a bill to fund public education in Virginia. John Trumbull completes *The Death of General Warren at the Battle of Bunker's Hill* in London.

1787   Charles Willson Peale's museum of art, natural history, and technology opens in Philadelphia. Joel Barlow publishes *The Vision of Columbus; A Poem in Nine Books*. Benjamin Rush argues for "female education" in an address to the Young Ladies' Academy of Philadelphia. John Trumbull completes *The Declaration of Independence, Philadelphia, 4 July 1776.* Royall Tyler's play *The Contrast* is performed at the John Street Theater in New York in April. Thomas Jefferson publishes his *Notes on the State of Virginia* in London.

1789   Noah Webster publishes his *Dissertations on the English Language: with Notes, Historical and Critical,* dedicated to Benjamin Franklin. David Ramsay publishes his *History of the American Revolution.*

1790   Noah Webster's *Collection of Essays and Fugitiv Writings* demonstrates his proposed orthographic reforms. President George Washington asks Congress to establish a national university. Judith Sargent Murray publishes "On the Equality of the Sexes" in the *Massachusetts Magazine, or, Monthly Museum of Knowledge and Rational Entertainment.*

1791   Robert Coram publishes his *Political Inquiries: to which is Added, a Plan for the General Establishment of Schools throughout the United States.* William Bartram publishes his *Travels through North and South Carolina, Georgia, East and West Florida, the Cherokee Country, the Extensive Territories of the Muscogulges or Creek Confederacy, and the Country of the Chactaws.*

1792   Hugh Henry Brackenridge publishes the first two volumes of *Modern Chivalry: Containing the Adventures of Captain John Farrago, and Teague O'Regan, His Servant.*

1793   Boston lifts a ban of more than forty years on theaters and the production of plays.

1794   Susanna Haswell Rowson publishes *Charlotte. A Tale of Truth* and performs in her *Slaves in Algiers; or, a Struggle for Freedom: A Play, interspersed with songs, in three acts* in Philadelphia. William Cobbett attacks Rowson and predicts social revolution in *A Kick for a Bite.* Timothy Dwight publishes *Greenfield Hill: A Poem in Seven Parts.*

1795–   Gilbert Stuart paints two busts and a full-length portrait of
1797   George Washington.

1797   Samuel Harrison Smith and Samuel Knox share the prize in a contest to design an American public school system, sponsored by the American Philosophical Society of Philadelphia. "Novel Reading, a Cause of Female Depravity" is published and widely printed in magazines and journals.

1798   Samuel Harrison Smith publishes his *Remarks on Education: Illustrating the Close Connection between Virtue and Wisdom.* Judith Sargent Murray publishes *The Gleaner* in three volumes.

1799   George Washington's Last Will and Testament bequeaths funds to found a national university. Samuel Knox publishes *An Essay on the Best System of Liberal Education, Adapted to the Genius of the Government of the United States.*

1800   Mason Locke Weems publishes his *Life of George Washington; with Curious Anecdotes, Equally Honourable to Himself and Exemplary to his Young Countrymen.* The U.S. government purchases Gilbert Stuart's portrait *George Washington* to display in the White House.

1803   President Thomas Jefferson invites William Bartram to serve as naturalist on a government-sponsored expedition up the Red River; Bartram declines due to poor health.

1804   John Vanderlyn paints *The Death of Jane McCrea.*

1805   Mercy Otis Warren publishes her *History of the Rise, Progress and Termination of the American Revolution.*

1806   Noah Webster publishes his *Compendious Dictionary.* President Thomas Jefferson repeats George Washington's request to Congress to establish a national university, with identical results.

Questions for Consideration

1. Why did the intellectuals of the Revolutionary generation view political divisions and cultural diversity as such a grave problem? Would an American language have solved the problem? Why or why not?

2. How would you characterize the republican reformers' educational vision? If you had been young in the 1790s, would you have wanted to attend a school of their design, or would you have preferred to learn at home with Margaretta? Would you have found the republican concept of moral education liberating or limiting? Is moral education for citizenship a legitimate goal of public schools?

3. Why did George Washington figure so prominently in the narratives of nationhood constructed by Revolutionary historians and artists? How did his image take shape in different hands and narratives? Did it serve different purposes for different audiences?

4. Why did the Revolutionary intellectuals so fear American popular culture? What meanings might early American sentimental novels and plays have held for their many readers and viewers?

5. Why couldn't intellectuals of the Revolutionary generation define American identity without also constructing an ethnic or racial Other? Can we do so today?

6. If the intellectuals of the Revolutionary generation had succeeded in creating a strong, unified, republican American culture between 1775 and 1800, how might the course of American history have been altered? Was this ever a real possibility? Why is their dream of American cultural unity and identity still alive today?

Selected Bibliography

INTRODUCTION

Bailyn, Bernard. *The Ideological Origins of the American Revolution.* Cambridge, Mass., 1967.

Berlin, Ira, and Ronald Hoffman, eds. *Slavery and Freedom in the Age of the American Revolution.* Charlottesville, Va., 1983.

Countryman, Edward. "'Out of the Bounds of Law': Northern Land Rioters in the Eighteenth Century," in *The American Revolution: Explorations in the History of American Radicalism,* edited by Alfred F. Young, 37–69. De Kalb, Ill., 1976.

Frey, Sylvia R. *Water from the Rock: Black Resistance in a Revolutionary Age.* Princeton, N.J., 1991.

Isaac, Rhys. *The Transformation of Virginia, 1740–1790.* Chapel Hill, N.C., 1982.

May, Henry F. *The Enlightenment in America.* New York, 1976.

Nash, Gary B. *The Urban Crucible: Social Change, Political Consciousness, and the Origins of the American Revolution.* Cambridge, Mass., 1979.

Norton, Mary Beth. *Liberty's Daughters: The Revolutionary Experience of American Women, 1750–1800.* Boston, 1980.

Pocock, J. G. A. *The Machiavellian Moment: Florentine Political Thought and the Atlantic Republican Tradition.* Princeton, N.J., 1975.

Royster, Charles. *A Revolutionary People at War: The Continental Army and American Character, 1775–1783.* Chapel Hill, N.C., 1979.

Shy, John. "The American Revolution: The Military Conflict Considered as a Revolutionary War," in *Essays on the American Revolution,* edited by Stephen G. Kurtz and James H. Hutson, 121–56. Chapel Hill, N.C., 1973.

Slaughter, Thomas P. *The Whiskey Rebellion: Frontier Epilogue to the American Revolution.* New York, 1986.

Weigley, Russell F. *The Partisan War: The South Carolina Campaign of 1780–1782.* Columbia, S.C., 1970.

Wood, Gordon S. *The Creation of the American Republic, 1776–1787.* Chapel Hill, N.C., 1969.

INVENTING AN AMERICAN LANGUAGE AND LITERATURE

Bloch, Ruth. *Visionary Republic: Millennial Themes in American Thought, 1756–1800.* Cambridge, Mass., 1985.

Cmiel, Kenneth. "'A Broad Fluid Language of Democracy': Discovering the American Idiom." *Journal of American History,* 79, no. 3 (1992): 913–36.

Elliott, Emory. *Revolutionary Writers: Literature and Authority in the New Republic, 1725–1810.* New York, 1982.

Silverman, Kenneth. *A Cultural History of the American Revolution.* New York, 1976.

Tuveson, Ernest Lee. *Redeemer Nation: The Idea of America's Millennial Role.* Chicago, 1968.

EDUCATING AMERICAN CITIZENS

Cremin, Lawrence A. *American Education: The National Experience, 1783–1876.* New York, 1980.

Hart, Sidney, and David C. Ward. "The Waning of an Enlightenment Ideal: Charles Willson Peale's Philadelphia Museum, 1790–1820." *Journal of the Early Republic,* 8 (1988): 389–418.

Kaestle, Carl F. *Pillars of the Republic: Common Schools and American Society, 1780–1860.* New York, 1983.

Kornfeld, Eve. "'Republican Machines' or Pestalozzian Bildung? Two Visions of Moral Education in the Early Republic." *Canadian Review of American Studies,* 20, no. 2 (Fall 1989): 157–72.

Yazawa, Melvin. *From Colonies to Commonwealth: Familial Ideology and the Beginnings of the American Republic.* Baltimore, 1985.

NARRATING NATIONHOOD

Cohen, Lester H. *The Revolutionary Histories: Contemporary Narratives of the American Revolution.* Ithaca, N.Y., 1980.

Harris, Neil. *The Artist in American Society: The Formative Years, 1790–1860.* Chicago, 1966.

Kornfeld, Eve. "From Republicanism to Liberalism: The Intellectual Journey of David Ramsay." *Journal of the Early Republic,* 9 (Fall 1989): 289–313.

Nochlin, Linda. *The Politics of Vision: Essays on Nineteenth-Century Art and Society.* New York, 1989.

Shaffer, Arthur H. *To Be an American: David Ramsay and the Making of the American Consciousness.* Columbia, S.C., 1991.

CONTESTING POPULAR CULTURE

Bloch, Ruth H. "The Gendered Meanings of Virtue in Revolutionary America." *Signs: Journal of Women in Culture and Society,* 13, no. 1 (1987): 37–58.

Davidson, Cathy N. *Revolution and the Word: The Rise of the Novel in America.* New York, 1986.

Kerber, Linda K. *Women of the Republic: Intellect and Ideology in Revolutionary America.* Chapel Hill, N.C., 1980.

Kornfeld, Eve. "Culture and Counterculture in Post-Revolutionary America." *Journal of American Culture,* 12, no. 4 (Winter 1989): 71–77.

Kornfeld, Eve. "Women in Post-Revolutionary American Culture: Susanna Haswell Rowson's American Career, 1793–1824." *Journal of American Culture,* 6, no. 4 (1983): 56–62.

Lewis, Jan. "The Republican Wife: Virtue and Seduction in the Early Republic." *William and Mary Quarterly,* 3rd ser., 44, no. 4 (1987): 689–721.

Watts, Steven. "Masks, Morals, and the Market: American Literature and Early Capitalist Culture, 1790–1820." *Journal of the Early Republic,* 6, (1986): 127–49.

ENCOUNTERING THE OTHER

Berkhofer, Robert F. *The White Man's Indian: Images of the American Indian from Columbus to the Present.* New York, 1978.

Bhabha, Homi K., ed. *Nation and Narration.* London, 1990.

Commager, Henry Steele, and Elmo Giodanetti, eds. *Was America a Mistake? An Eighteenth-Century Controversy.* New York, 1967.

Jennings, Francis. *The Invasion of America: Indians, Colonialism, and the Cant of Conquest.* Chapel Hill, N.C., 1975.

Kornfeld, Eve. "Encountering 'the Other': American Intellectuals and Indians in the 1790s." *William and Mary Quarterly,* 3rd ser., 52, no. 2 (April 1995): 285–314.

Minh-ha, Trinh T. *Woman, Native, Other: Writing Postcoloniality and Feminism.* Bloomington, Ind., 1989.

Pearce, Roy Harvey. *The Savages of America: A Study of the Indian and the Idea of Civilization,* rev. ed. Baltimore, 1965.

Pratt, Mary Louise. *Imperial Eyes: Travel Writing and Transculturation.* London, 1992.

Regis, Pamela. *Describing Early America: Bartram, Jefferson, Crèvecoeur, and the Rhetoric of Natural History.* De Kalb, Ill., 1992.

Said, Edward W. *Culture and Imperialism.* New York, 1993.

Index

Printed in the United States
By Bookmasters